W9-APQ-996

Chromatin

The Practical Approach Series

SERIES EDITORS

D. RICKWOOD
Department of Bio Biology, University of Essex, Wivenhoe Park, Colchester, Essex CO4 3SQ, UK

B. D. HAMES
Department of Biochemistry and Molecular Biology University of Leeds, Leeds LS2 9JT, UK

See also the Practical Approach web site at http://www.oup.co.uk/PAS

★ indicates new and forthcoming titles

Affinity Chromatography
★ Affinity Separations
Anaerobic Microbiology
Animal Cell Culture (2nd edition)
Animal Virus Pathogenesis
Antibodies I and II
Antibody Engineering
★ Antisense Technology
★ Applied Microbial Physiology
Basic Cell Culture
Behavioural Neuroscience
Bioenergetics
Biological Data Analysis
Biomechanics—Materials
Biomechanics—Structures and Systems
Biosensors
Carbohydrate Analysis (2nd edition)
Cell-Cell Interactions
The Cell Cycle
Cell Growth and Apoptosis
Cellular Calcium
Cellular Interactions in Development
Cellular Neurobiology
★ Chromatin
Clinical Immunology
★ Complement
Crystallization of Nucleic Acids and Proteins
Cytokines (2nd edition)
The Cytoskeleton
Diagnostic Molecular Pathology I and II
DNA and Protein Sequence Analysis
DNA Cloning 1: Core Techniques (2nd edition)
DNA Cloning 2: Expression Systems (2nd edition)
DNA Cloning 3: Complex Genomes (2nd edition)
DNA Cloning 4: Mammalian Systems (2nd edition)
★ Drosophila (2nd edition)

Electron Microscopy in Biology
Electron Microscopy in Molecular Biology
Electrophysiology
Enzyme Assays
Epithelial Cell Culture
Essential Developmental Biology
Essential Molecular Biology I and II
Experimental Neuroanatomy
Extracellular Matrix
Flow Cytometry (2nd edition)
Free Radicals
Gas Chromatography
Gel Electrophoresis of Nucleic Acids (2nd edition)
Gel Electrophoresis of Proteins (2nd edition)
Gene Probes 1 and 2
Gene Targeting
Gene Transcription
★ Genome Mapping
Glycobiology
Growth Factors
Haemopoiesis
Histocompatibility Testing
HIV Volumes 1 and 2
★ HPLC of Macromolecules (2nd edition)
Human Cytogenetics I and II (2nd edition)
Human Genetic Disease Analysis
★ Immunochemistry 1
★ Immunochemistry 2
Immunocytochemistry
In Situ Hybridization
Iodinated Density Gradient Media
Ion Channels
Lipid Modification of Proteins
Lipoprotein Analysis
Liposomes
Mammalian Cell Biotechnology
Medical Parasitology
Medical Virology
★ MHC Volumes 1 and 2
★ Molecular Genetic Analysis of Populations (2nd edition)
Molecular Genetics of Yeast
Molecular Imaging in Neuroscience
Molecular Neurobiology
Molecular Plant Pathology I and II
Molecular Virology
Monitoring Neuronal Activity
Mutagenicity Testing
★ Mutation Detection
Neural Cell Culture
Neural Transplantation
Neurochemistry (2nd edition)
Neuronal Cell Lines
NMR of Biological Macromolecules
Non-isotopic Methods in Molecular Biology
Nucleic Acid Hybridisation
Oligonucleotides and Analogues
Oligonucleotide Synthesis

PCR 1
PCR 2
★ PCR 3: PCR In Situ Hybridization
Peptide Antigens
Photosynthesis: Energy Transduction
Plant Cell Biology
Plant Cell Culture (2nd edition)
Plant Molecular Biology
Plasmids (2nd edition)
★ Platelets
Postimplantation Mammalian Embryos
Preparative Centrifugation
Protein Blotting
Protein Engineering
★ Protein Function (2nd edition)
Protein Phosphorylation
Protein Purification Applications
Protein Purification Methods
Protein Sequencing
★ Protein Structure (2nd edition)
Protein Structure Prediction
Protein Targeting
Proteolytic Enzymes
Pulsed Field Gel Electrophoresis
RNA Processing I and II
★ RNA-Protein Interactions
★ Signalling by Inositides
Subcellular Fractionation
Signal Transduction
Transcription Factors
Tumour Immunobiology

Chromatin

A Practical Approach

Edited by

H. GOULD

The Randall Institute

King's College London

Oxford New York Tokyo

OXFORD UNIVERSITY PRESS

1998

Oxford University Press, Great Clarendon Street, Oxford OX2 6DP

Oxford New York
Athens Auckland Bangkok Bogota Bombay Buenos Aires Calcutta
Cape Town Chennai Dar es Salaam Delhi Florence Hong Kong Istanbul
Karachi Kuala Lumpur Madrid Melbourne Mexico City Mumbai
Nairobi Paris São Paolo Singapore Taipei Tokyo Toronto Warsaw

and associated companies in
Berlin Ibadan

Oxford is a trade mark of Oxford University Press

Published in the United States
by Oxford University Press Inc., New York

A catalogue record for this book is available from the British Library

Library of Congress Cataloging in Publication Data
Chromatin : a practical approach / edited by H. Gould.
(Practical approach series)
Includes bibliographical references and index.
1. Chromatin—Laboratory manuals. 2. Chromatin—Research —Methodology. I. Gould, H. (Hannah) II. Series.
QH599.C45 1998 572.8'7—dc21 97-44752

ISBN 0 19 963599 4 (Hbk)
0 19 963598 6 (Pbk)

Typeset by Footnote Graphics, Warminster, Wilts
Printed in Great Britain by Information Press Ltd, Eynsham, Oxon.

Contents

9. Transcriptional and structural analysis of chromatin assembled *in vitro* 173

Michael J. Pazin and James T. Kadonaga

10. Assembly of chromatin and nuclear structures in *Xenopus* egg extracts 195

Geneviève Almouzni

Contributors

JAMES ALLAN
Department of Biochemistry, University of Edinburgh, George Square, Edinburgh EH8 9XD, UK.

GENEVIÈVE ALMOUZNI
Institut Curie, Section de recherche, UMR 144, Dynamique de la chromatine, 26, rue d'Ulm-75231, Paris Cedex 05, France.

PETER D. CARY
Biophysics Laboratories, University of Portsmouth School of Biological Sciences, St Michael's Building, White Swan Road, Portsmouth PO1 2DT, UK.

DAVID J. CLARK
Laboratory of Cellular and Developmental Biology, NIDDK, Building 6, Room B1–12, National Institute of Health, Bethesda, MD 20892–2715, USA.

COLYN CRANE-ROBINSON
Biophysics Laboratories, University of Portsmouth School of Biological Sciences, St Michael's Building, White Swan Road, Portsmouth PO1 2DT, UK.

COLIN DAVEY
Department of Biochemistry, University of Edinburgh, George Square, Edinburgh EH8 9XD, UK.

SARAH C. R. ELGIN
Department of Biology, Box 1229, Washington University, St Louis, MO 63130, USA.

SUSAN M. GASSER
Swiss Institute for Experimental Cancer Research, Chemin des Boveresses 155, CH-1066 Epalinges s/Lausanne, Switzerland.

JAMES T. KADONAGA
Department of Biology, 0347, and Center for Molecular Genetics, Pacific Hall, Room 2216, University of California, San Diego, 9500 Gilman Drive, La Jolla, CA 92093–0347, USA.

MICHAEL J. PAZIN
Department of Biology, 0347, and Center for Molecular Genetics, Pacific Hall, Room 2216, University of California at San Diego, 9500 Gilman Drive, La Jolla, CA 92093–0347, USA. Current address: Cutaneous Biology Research Center, Massachusetts General Hospital, 13th Street, Building 149, Charlestown, MA 02129, USA.

SARI PENNINGS
Department of Biochemistry, University of Edinburgh, George Square, Edinburgh EH8 9XD, UK.

GERD P. PFEIFER
Biology Department, Beckman Research Institute of the City of Hope, City of Hope Medical Center, Duarte, CA 91010, USA.

ARTHUR D. RIGGS
Biology Department, Beckman Research Institute of the City of Hope, City of Hope Medical Center, Duarte, CA 91010, USA.

JUDITH SINGER-SAM
Biology Department, Beckman Research Institute of the City of Hope, City of Hope Medical Center, Duarte, CA 91010, USA.

MARCI J. SWEDE
Vyrex Corporation, 2159 Avendia de la Playa, La Jolla, CA 92037, USA.

FRITZ THOMA
Institut für Zellbiologie, Eidgenössische Technische Hochschule, ETH-Hönggerberg, CH-8093 Zürich, Switzerland.

JEAN O. THOMAS
Department of Biochemistry, University of Cambridge, 80 Tennis Court Road, Cambridge CB2 1GA, UK.

ALAN W. THORNE
Biophysics Laboratories, University of Portsmouth School of Biological Sciences, St Michael's Building, White Swan Road, Portsmouth PO1 2DT, UK.

YEGOR S. VASSETZKY
Swiss Institute for Experimental Cancer Research, Chemin des Boveresses 155, CH-1066 Epalinges s/Lausanne, Switzerland.

LORI L. WALLRATH
Department of Biochemistry, 4-772 Bowen Science Building, University of Iowa, Iowa City, IA 52242, USA.

Abbreviations

A+T	adenine + thymine
A_{260}	absorbance at 260 nm
ASW	artificial sea water
ATZ	anilinothiazolinone
bp	base pair
BSA	bovine serum albumin
CAT	chloramphenicol acetyl transferase
Ci	Curie
CSF	cytostatic factor
DAPI	diaminidophenylindol
DHCC	3,3′-dihexyloxacarbocyanine
Δlk	change in linking number
DMS	dimethyl sulfate
DNase I	deoxyribonuclease I
dNTP	deoxyribonucleoside triphosphate
DTT	dithiothreitol
EDTA	ethylenediamine-tetraacetic acid
EGTA	ethyleneglycol tetraaceticacid
EM	electron microscopy
ESMS	electrospray ionization mass spectrometry
EtBr	ethidium bromide
Exo III	exonuclease III
G+C	guanine + cytosine
HCG	human chorionic gonadotropin
HEMG	Hepes–EDTA–magnesium–glycerol
HMG	high-mobility group
HSE	high-speed extract
hsp	heat-shock protein
IAA	isoamyl alcohol
LIS	lithium 3′,5′-diiodosalicylate
LMPCR	ligation-mediated PCR
LSE	low-speed extract
MALDITOF	matrix assisted laser desorption ionization time of flight (mass spectrometry)
MAR	matrix associated region
MEM	minimum essential medium
MNase	micrococcal nuclease
MPE–Fe(II)	methidium propyl-EDTA-iron (II)
MPF	maturation (or mitosis) promoting factor
MS/MS	tandem mass spectrometry

NHCP	non-histone protein
NIM	nuclear isolation media
NMPCL	nuclear matrix-pore, complex-lamina
NP-40	Nonidet P-40
PBS	phosphate-buffered saline
PCNA	proliferating cellular nuclear antigen
PCR	polymerase chain reaction
PEG	polyethylene glycol
PMSF	phenylmethylsulfonyl fluoride
PMSG	pregnant mare serum gonadotropin
PNK	polynucleotide kinase
ppNa	sodium pyrophosphate
PTH	phenylthiohydantoin
PVDF	polyvinylidene difluoride
PvOH	polyvinyl alcohol
RNase	ribonuclease
RNAse A	ribonuclease A
rNTP	ribonucleoside triphosphate
RPA	replication protein A
RT	reverse transcriptase
SAR	scaffold-attachment region
SDS	sodium dodecyl sulfate
SMF	scanning force microscopy
SNF	soluble nuclear fraction
SSC	standard saline citrate
ssDNA	single-stranded DNA
TAE	Tris–acetate–EDTA
TBE	Tris–borate–EDTA
TCA	trichloroacetic acid
TE	Tris–EDTA
TEM	transmission electron microscope
TEMED	*N*,*N*,*N*′,*N*′-tetramethylethylenediamine
TEP	Tris–EDTA–PMSF
TF	transcription factor
TFA	trifluoracetic acid
TMS	Tris–magnesium–sucrose
TNM	Tris–Na–magnesium
TPE	Tris–phosphate–EDTA

Wood block engraving by Hokusai. © The British Museum.

Introduction

H. GOULD

What is true of an elephant is not altogether true of *E. coli*

An estimated 1.5 billion years ago, a prokaryotic cell founded a line that gave rise to eukaryotes. The designation eukaryote, conferred on the descendants of this cell by Edouard Chatton in 1937, stems from their most prominent feature, the cell nucleus. Modern techniques have shown us that the DNA is also organized into a higher-order structure, formed by a series of octameric histone complexes spaced at *c.* 200 base-pair intervals. The processes of replication and transcription, inherited from prokaryotes, require the chromatin structure to unfold. Genes that are active at any given time are in the unfolded fraction. The population of unfolded genes varies with the stage of development of the organism, the cell lineage, the stage in the cell cycle, and the local conditions, such as the phenotypes and spatial arrangement of neighbouring cells, the local concentrations of nutrients and hormones, and the temperature (as in heat shock). Miraculously, chromatin can read these signals or, if the fate of the cell is 'determined', disregard them.

A venerable misconception in biology was the belief that the role of histones is passive, that non-histone proteins carry all the instructions for the activity of chromatin, and that changes in gene expression are dictated by the *de novo* synthesis, post-translational modification (e.g. phosphorylation or dephosphorylation) and/or compartmentation of these proteins. This view is still voiced in some quarters, but has become overtaken by the evidence for the participation of histones in the dynamic remodelling of chromatin during gene activation and repression. The early fibre diffraction studies on DNA yielded averaged dimensions for the double helix. Much later it was discovered that the twist, bend, and roll of the bases varies locally with sequence. Such variations are recognized as much by the histone octamers as by DNA-binding non-histone proteins, so that there is a considerable variation in the affinity of the octamers for different DNA sequences. This selectivity is one of the factors that determines the positioning of nucleosomes on the regulatory sequences, e.g. promoters and enhancers, of many genes. Competition with non-histone proteins that recognize sequences engaged in nucleosome structure may change the conformation or position of the nucleosome, or even push it off the site altogether. In some documented cases such displacement requires the expenditure of energy (by ATP hydrolysis) or the covalent modification (acetylation) of the histones to weaken their hold on the DNA. Is it the non-histone proteins or the histone octamers that read the DNA?

The answer is both, for chromatin structure, as well as the sequences of RNA and of proteins, are encoded by DNA, and the former is employed to regulate expression of the latter.

We begin this manual with a chapter on the 'Isolation and fractionation of chromatin and linker histones' (Chapter 1) by Jean Thomas and 'Extraction and separation of core histones and non-histone chromosomal proteins' (Chapter 2) by Alan Thorne, Peter Cary and Colyn Crane-Robinson. One of the manifestations of gene activation is the increased accessibility of regulatory sequences (promoters, enhancers, locus control regions) extending over >200 bp, resulting from the replacement of histone octamers by non-histone DNA-binding proteins and accessory factors. Thus, to zoom in on these sites one can use a variety of probes (most often DNA cleavage enzymes or chemical reagents) for DNA accessibility. The widely used probe discussed by Lori Wallrath, Marci Swede, and Sally Elgin in 'Mapping chromatin structure in *Drosophila*' (Chapter 3) is DNase I. However, the elegant technique of indirect labelling, invented in Elgin's laboratory to map the cutting sites, revealed as a series of strong bands in a DNA sequencing gel, can be applied in combination with other cleavage strategies.

The image of the bound non-histones proteins is similarly revealed by '*In vivo* footprint and chromatin analysis by LMPCR' (Chapter 4), as described in full detail by Arthur Riggs, Judith Singer-Sam, and Gerd Pfeifer. Nucleosome arrays are wound into a higher-order structure and these 30 nm thick threads are looped out of a nuclear scaffold. Susan Gasser and Yegor Vassetzky disclose in 'Analysis of nuclear scaffold attachment regions' (Chapter 5) how to map the attachment sites in the sequence of DNA and discuss their possible significance. Our knowledge of the gross structure of nucleosomes and higher-order chromatin structure derives from electron microscopy. How to assess the gross structure of chromatin by this means is described in 'Electron microscopy of chromatin' (Chapter 6) by Fritz Thoma. Electron microscopy was indispensable in the early studies of chromatin structure and will become increasingly important with the development of more faithful methods of chromatin reconstitution (see Chapter 9), as well as new ways of viewing these structures in the electron microscope with minimal perturbation. The winding of DNA around the histone octamer and formation of higher-order structures generate the topological properties of chromatin described by David Clark in 'Counting nucleosome cores on circular DNA using topoisomerase I' (Chapter 7). The techniques elaborated in this chapter reappear in Chapter 9 as a means of characterizing chromatin templates reconstituted for *in vitro* transcription.

The last three chapters all deal with chromatin reconstitution, the full power of which has yet to emerge. Here we see the application of all the previously described methods. Colin Davey, Sari Pennings, and Jim Allan are concerned with '*In vitro* reconstitution and analysis of nucleosome positioning' (Chapter 8) and Michael Pazin and Jim Kadonaga with 'Transcriptional and structural analysis of chromatin assembled *in vitro*' (Chapter 9). In the final chapter, 'Assembly of chromatin and nuclear structures in *Xenopus* egg extracts', Geneviève Almouzni considers the functional reconstitution of whole nuclei, which perform both transcription and replication.

My choice of topics in this book reflects the complementarity of *in vivo* and *in vitro* analyses and functional reconstitution methods in understanding mechanisms of gene expression, encompassed by the interplay of non-histone proteins and chromatin templates. It has been fascinating to witness the discoveries of the last 5–10 years. It appears that no two genes share a single mode of regulation, and even the same gene in different tissues may invoke different players. Some general mechanisms, such as 'promotion', 'enhancement', 'locus control', 'silencing', and 'insulation', exhibited by promoters, enhancers, locus control elements, silencers, and insulators respectively, can be distinguished, but these operate through different sets of proteins and chromatin substrates in every situation. Scientists from many fields of biology, genetics, developmental biology, immunology, and physiology, to mention only a few, can scarcely now escape an interest in the mechanisms by which genes are selectively expressed, and therefore the role of chromatin structure in these mechanisms. Mutations in signalling proteins that cause the misreading of cellular signals by chromatin lead to developmental defects and certain types of cancer. This shows that chromatin methods have a direct bearing on medical research.

A companion volume, *Chromatin Structure and Gene Expression* (ed. S.C.R. Elgin, IRL at Oxford University Press, 1995), provides an excellent introduction to the conceptual framework in chromatin research and presents classic examples of the application of chromatin methodology. The present volume attempts to demystify working with chromatin, and so facilitate wider use of the methods. Many more hands will be required to produce the blueprint for an elephant.

1

Isolation and fractionation of chromatin and linker histones

JEAN O. THOMAS

1. Introduction

Nuclei from different sources usually have a common pattern of core histones (components of the histone octamer) as revealed by SDS/polyacrylamide gel electrophoresis, but in some cases they have distinctive linker histone variants that may correlate with particular states of cell differentiation and transcriptional activity (more subtle H1 subtypes present in many cell types are not resolved in SDS gels). Nuclei from different sources may also differ in nucleosome repeat length. Procedures are described for the isolation of nuclei from rat liver, and from chicken erythrocytes and sea-urchin sperm which contain extreme H1 variants and have repeat lengths longer than the canonical 200 base pairs of rat liver and many other tissues. A method is also described for the isolation of nuclear populations from various mammalian brain cell types; the distinctive feature here is that neurones from the cerebral cortex have a shorter repeat length than the glial cells from the same tissue or cerebellar neurones, and only half the H1 content.

Soluble chromatin fragments, of particular size ranges, may be produced under mild conditions by controlled enzymic fragmentation of nuclear chromatin and size-fractionation. These are useful for studying the bulk properties of chromatin (e.g. hydrodynamic properties under various conditions and the effects of removal or substitution of linker histones) as well as in studies of the properties of localized regions of the genome, identified by Southern blotting of the DNA. The nuclear attachments of the chromatin are, of course, lost during the preparation of soluble fragments, and under commonly used (hypotonic) conditions for the release of nuclear chromatin fragments the higher-order structure (30 nm filament) is unfolded, although at a gross level at least this can apparently be re-formed by manipulation of the ionic strength.

Procedures are also described for removal of the linker histones under relatively mild conditions to give H1-depleted chromatin (which can be reconstituted with a different H1 variant or with a modified H1, for example, if necessary), as well as for the isolation of linker histones and for their characterization (e.g. with respect to phosphorylation state). Methods for the isolation of core histones are described in Chapter 2; see also (1).

Methods for the isolation and fractionation of chromatin have also been described and discussed elsewhere (2).

2. Isolation of nuclei

2.1 Rat liver nuclei

The method described is essentially that of Hewish and Burgoyne (3). The nuclear isolation buffer contains 0.34 M sucrose (isotonic for nuclei), spermine and spermidine to stabilize nuclei by ensuring the condensation of the DNA, and EDTA and EGTA to inhibit the action of endogenous divalent cation-dependent endonucleases. The inclusion of KCl (60 mM) and NaCl (15 mM) aims to give an ionic strength close to that inside the nucleus *in vivo* (3). Homogenization of the liver in buffer containing 0.34 M sucrose disrupts the plasma membrane and releases nuclei. A crude nuclear pellet is collected by centrifugation, resuspended in 2.1 M sucrose, and centrifuged at high speed. A relatively pure nuclear pellet is obtained; very little else will sediment through 2.1 M sucrose.

Protocol 1. Isolation of rat liver nuclei

Equipment and reagents

- Adult rats (2–3 months)
- Homogenizer (glass homogenizer with two plungers: a motor-driven, tight-fitting Teflon plunger and a hand-held Perspex plunger)
- Muslin (cheesecloth)
- 30 ml Corex glass centrifuge tubes and 50 ml plastic centrifuge tubes
- Fixed-angle rotor (e.g. Sorvall SS-34)
- Swing-out ultracentrifuge rotor (e.g. Beckman SW28 rotor or equivalent) and 17 ml or 38 ml polyallomer tubes
- 0.34 M sucrose–buffer A, pH 7.5 (nuclear wash buffer). Buffer A: 15 mM Tris–HCl, pH 7.5, 15 mM NaCl, 60 mM KCl, 0.5 mM spermidine hydrochloride (from frozen 50 mM stock, Sigma, cat. no. S-2501), 0.15 mM spermine hydrochloride (from frozen 50 mM stock, Sigma, cat. no. S-2876), 15 mM 2-mercaptoethanol (final pH ~7.5)
- 0.34 M sucrose–buffer A^+ (nuclear isolation buffer): 0.34 M sucrose–buffer A plus 2 mM Na_2EDTA, pH 7.0, 0.5 mM Na_2EGTA, pH 7.0
- 2.1 M sucrose–buffer A^+: 2.1 M sucrose–buffer A plus 2 mM Na_2EDTA, pH 7.0, 0.5 mM Na_2EGTA, pH 7.0

Method

All operations should be carried out at 0–4 °C. For a fairly large scale preparation use three rats.

1. Remove the livers (~15 g/adult rat) from three rats killed by an approved procedure. Transfer immediately to ice-cold 0.34 M sucrose–buffer A^+ to wash. Alternatively, use frozen livers provided they are frozen in liquid nitrogen immediately after dissection (e.g. from animals which are being sacrificed for other purposes; this minimizes animal usage) and then stored at – 80 °C.

2. Mince the fresh liver with scissors in a beaker with at most a few ml of 0.34 M sucrose–buffer A^+ and then homogenize thoroughly, mechanically and then manually, until dispersed. Do this in the cold-room, or at least keep the homogenizer in ice. If frozen livers are used, break them into pieces before adding to the buffer to facilitate rapid thawing.
3. Filter the homogenate through four layers of muslin previously washed in 0.34 M sucrose–buffer A^+. After filtering, wash the muslin thoroughly several times with the same buffer to give ~120 ml of homogenate for a preparation starting with three livers.
4. Layer onto ~10 ml cushions of a 1:1 mixture of 0.34 M sucrose–buffer A^+ and 2.1 M sucrose–buffer A^+ in 50 ml plastic centrifuge tubes.
5. Centrifuge for 15 min at 12 000 g_{max} in a fixed-angle rotor (e.g. Sorvall SS-34) in a refrigerated centrifuge. (Be there when the centrifuge stops!).
6. Decant the supernatant carefully, keeping the (loose) pellet down. Stop pouring when the pellet starts to move.
7. Resuspend the reddish-coloured pellet in 2.1 M sucrose–buffer A^+, first by vortex-mixing and then with a 10 ml pipette, to a final volume of ~85 ml for three livers.
8. For three livers, layer in 14 ml portions onto 3 ml cushions of 2.1 M sucrose–buffer A^+ in (e.g.) six 17 ml Beckman SW28 tubes. For twice the scale, layer onto 6–7 ml cushions in (e.g.) 38 ml SW28 tubes. Tubes must be filled to within ~2 mm of the top.
9. Centrifuge for 50 min at 100 000 g_{av} to pellet the nuclei. Scrape out the top layer of lipid with a spatula and decant the supernatant. Drain the tube and wipe the inside with paper tissue to remove remaining traces of lipid and supernatant.
10. Wash the nuclei by resuspending the pellet in 0.34 M sucrose–buffer A (**NB** *No* EDTA, EGTA) by gentle pipetting. Transfer to one or two 30 ml Corex glass tubes and centrifuge at 2000 g_{max} for 5 min. Repeat the resuspension and centrifugation steps. The nuclei at this point should look white (or creamy white). Check with a phase-contrast microscope that they look clean and intact.
11. Resuspend in 0.34 M sucrose–buffer A. Read the A_{260} (e.g. of 10 μl) in 1 M NaOH and adjust the volume to give A_{260} ~50. This is very approximately (and ignoring RNA) 2.5 mg DNA/ml (assuming A_{260} = 20 for 1 mg/ml DNA), or ~5 mg chromatin/ml (assuming histone:DNA ~1:1 (w/w)). Expect ~7 ml of nuclei at A_{260} = 50 from one 15 g liver.

2.2 Chicken erythrocyte nuclei

The mature circulating red blood cells of birds, fish, reptiles, and amphibia have retained their nuclei and these may be readily isolated. The nuclei, which are smaller and more condensed than those from rat liver, are transcriptionally inert and contain an extreme H1 variant, H5, which largely, but not completely, replaces H1 during erythropoiesis. The procedure for the isolation of erythrocytes and nuclei described below is based on that of Weintraub *et al.* (4). Washed erythrocytes may be conveniently stored for months at – 80 °C if they are needed only for the isolation of nuclei (they lyse on thawing).

2.2.1 Chicken erythrocytes

Protocol 2. Isolation and storage of chicken erythrocytes

Equipment and reagents

- 500 ml, wide-necked, plastic centrifuge bottles
- Muslin (cheesecloth)
- Collection buffer: 3.4% (w/v) tri-sodium citrate (200 ml buffer/litre of blood), autoclaved and containing 0.1% (w/v) chloramphenicol which should be added just before collection of the blood.
- Isolation buffer (SSC; standard saline citrate): 10 mM Tris–HCl, pH 7.5, 0.14 M NaCl, 15 mM tri-sodium citrate, and containing 0.25 mM PMSF added from a stock solution (50 mM in propan-2-ol) immediately before erythrocyte isolation. SSC is conveniently kept as a 10 × stock.
- Stoppered tissue-culture tubes

Method

The following is for 1 litre of pooled chicken blood, which gives a good supply of storable erythrocytes. All operations should be carried out at 0–4 °C.

1. Use fresh blood , preferably collected straight from the chickens into cold collection buffer. Immediately, mix the blood and buffer well to prevent clotting. Keep on ice and isolate the red cells as soon as possible (within an hour or so of sacrifice). If the blood is obtained from a commercial chicken factory, make sure that it is collected and mixed with ice-cooled buffer as soon as possible after slaughter, and transported to the laboratory on ice.
2. Filter the blood through one layer of muslin to remove any clots, etc.
3. Divide the blood between four 500 ml, wide-necked, plastic centrifuge bottles (~ 250 ml per bottle). Fill the bottle up with SSC, mix well, and centrifuge at 1500 g_{max} for 3 min (e.g. in a Sorvall GS.3 rotor).
4. Lift the bottles carefully out of the centrifuge without disturbing the pellets, and remove the supernatant by aspiration with a sawn-off 10 ml pipette attached to a water pump. The pellet of red cells is covered with a thin cream-coloured layer of white cells: the 'buffy coat'.
5. Remove as much as possible of the remaining supernatant and buffy coat by aspiration without disturbing the red cell pellet.
6. Swirl the bottle to resuspend the red cells, add more SSC, and repeat the centrifugation and aspiration steps.

7. Repeat the washing and aspiration steps until no more white cells are visible (usually two or three times more), meanwhile combining the red cells into two bottles and then into one.
8. Resuspend the concentrated red cells by swirling, then divide into appropriate aliquots in stoppered tissue-culture tubes (e.g. in 1, 5, and 10 ml aliquots), and freeze in dry ice. Store at – 80 °C.

2.2.2 Chicken erythrocyte nuclei

The procedure described is for erythrocytes that have been stored at – 80 °C (*Protocol 2*). It involves lysis of the cells during thawing in 0.34 M sucrose–buffer A (*Protocol 1*) in the presence of the non-ionic detergent Nonidet P-40 (NP-40). It works equally well for fresh, washed, erythrocytes, but lysis with NP-40 should be monitored under a phase-contrast microscope to make sure that it is complete and that the nuclei are not trapped inside empty 'bags' of plasma membrane. If this happens the yield of soluble chromatin later will be very low.

For lysis of the cells, 0.34 M sucrose–buffer A may be replaced with 0.34 M sucrose, 10 mM Tris–HCl, 3 mM $MgCl_2$ (TMS), the Mg^{2+} serving to stabilize the nuclei in the absence of spermine/spermidine. For nuclei from other sources there is a risk that Mg^{2+} will activate endogenous nucleases, but chicken erythrocyte nuclei appear to be relatively free of this problem.

Protocol 3. Isolation of chicken erythrocyte nuclei

Equipment and reagents

- 0.34 M sucrose–buffer A (*Protocol 1*) or TMS buffer (TMS buffer: 10 mM Tris–HCl, 3 mM $MgCl_2$, 0.34 M sucrose, pH 7.5), both with and without 0.5% NP-40
- Muslin
- Centrifuge tubes (30 ml Corex glass or 50 ml plastic)

Method

1. Thaw the frozen cells (10 ml of frozen cells/~ 150 ml of buffer) in 0.34 M sucrose–buffer A (or TMS buffer) containing 0.5% NP-40 at 37 °C, which causes lysis. For complete lysis make sure that the cells actually reach 37 °C, not just that they thaw out.
2. Filter through two layers of muslin into an ice-cooled flask.
3. Collect the nuclei by centrifugation at 2000 g_{max} for 4 min. Discard the supernatant.
4. Wash the nuclear pellets in the same buffer, but without NP-40, by resuspending gently (low-speed vortex mixing). Centrifuge again.
5. Continue washing and centrifuging, usually three times, until the supernatant is no longer red.

Protocol 3. *Continued*

6. Resuspend the nuclei in 0.34 M sucrose–buffer A, 0.25 mM PMSF. Read the A_{260} (e.g. of 10 μl) in 1 M NaOH and adjust the volume to give $A_{260} \sim 50$ (i.e. ~ 5 mg chromatin/ml). Expect ~ 400 A_{260} units of nuclei per ml of packed erythrocytes.

2.3 Sea-urchin sperm nuclei

The procedure described below works well for the two species of sea-urchin easily obtainable in the UK (from the University Marine Biology Station, Millport, Isle of Cumbrae, Scotland): *Echinus esculentus* (large, pink/purple; breeding season February–April depending on sea temperature; good sperm are produced only within this period) and *Psammechinus miliaris* (smaller, dark green; littoral; breeding season roughly March–June).

2.3.1 Sea-urchin sperm

Spermatozoa are collected when the urchins are mature and can be stored frozen in artificial sea water/glycerol at – 80 °C for at least a year.

Protocol 4. Isolation of sea-urchin sperm

Equipment and reagents

- *E. esculentus* or *P. miliaris* (e.g. 12, since males and females look identical)
- Muslin
- Artificial sea water (ASW): 0.42 M NaCl, 9 mM KCl, 9.3 mM $CaCl_2$, 22.9 mM $MgCl_2$, 25.5 mM $MgSO_4$, 2.15 mM $NaHCO_3$

Method

1. Determine the sex of the urchins by placing each of them, mouth upwards, on a glass beaker filled to the brim with artificial sea water (ASW) and standing in a tray of ice. Inject ~ 5 ml of 0.5 M KCl into the soft tissue around the mouth; a 50 × 0.9 mm needle is convenient. The urchin will release either spermatozoa or eggs into the beaker; these are easily distinguishable.
2. Leave the males on beakers on ice and, if necessary, inject with more KCl. They will continue to produce sperm for a few hours. Once production stops, or slows down significantly, and the water begins to turn yellow/green, remove the urchin and pour off the ASW from the whitish settled sperm.
3. Resuspend the sperm in ice-cold ASW, filter through two layers of muslin and collect by centrifugation at 2000 g_{max} for 4 min.
4. Wash the sperm twice with cold ASW by resuspension/centrifugation

and then resuspend the loose pellet in 40% ASW/60% glycerol (v/v) to give a dense suspension.

5. Freeze in 1 ml or 5 ml aliquots in capped tissue-culture tubes in dry ice and store at − 80 °C.

2.3.2 Sea-urchin sperm nuclei

We isolate sperm nuclei (5) using a slight modification of the method of Spadafora *et al.* (6). The tails are detached from the sperm heads by homogenization in non-ionic detergent. This is carried out in sucrose/Mg^{2+} to stabilize the nuclei, which are released as the membrane surrounding the sperm head is simultaneously disrupted. Sperm (and spermatid) nuclei may be further purified in Percoll gradients (7) if required.

Protocol 5. Isolation of sea-urchin sperm nuclei

Equipment and reagents

- Glass homogenizer with tight-fitting plunger
- Homogenization buffer: 0.25 M sucrose, 50 mM Tris–HCl, pH 7.5, 2 mM $MgCl_2$, 1% (v/v) Triton X-100
- Wash buffer: as homogenization buffer *but without Triton X-100*
- Sucrose cushion: 2 M sucrose, 50 mM Tris–HCl, pH 7.5, 2 mM $MgCl_2$
- 0.34 M sucrose–buffer A (*Protocol 1*)

Method

Carry out all operations at 0–4 °C.

1. Thaw the frozen sperm directly into homogenization buffer (e.g. a 5 ml aliquot of sperm in 40% ASW/60% glycerol into 10 ml of buffer).
2. Homogenize with 10 up-and-down strokes in a tight-fitting glass homogenizer to detach the tails from the sperm heads.
3. Collect the nuclei by centrifugation at 2000 g_{max} for 6 min. Discard the supernatant, which contains mainly sperm tails.
4. Wash the nuclei three times with wash buffer. (The nuclei sometimes clump a little at this stage, but can still be satisfactorily digested to give chromatin.)
5. Examine the nuclei under the light microscope. They are usually quite clean, but if there is appreciable contamination with tails they can be further purified by centrifugation through a 5 ml cushion of 2 M sucrose, 50 mM Tris–HCl, pH 7.5, 2 mM $MgCl_2$ at 10 000 g_{max} for 20 min, leaving the tails behind.
6. Resuspend the nuclei in 0.34 M sucrose–buffer A (*Protocol 1*) to give $A_{260} \sim 50$ (read in 1 M NaOH). It will be necessary to incubate the nuclei with a small amount of micrococcal nuclease and 1 mM $CaCl_2$ before taking the reading, because the undigested sperm chromatin is insoluble or the undigested sperm nuclei do not lyse.

2.3.3 Sea-urchin spermatid nuclei

E. esculentus mature in early February and *P. miliaris* in March so they should be harvested for spermatids about a month before this. The spermatid nuclei have to be extracted directly from the gonads (8) which, like the animals, look superficially the same irrespective of sex. Spermatid nuclei may be further purified to remove contaminating sperm nuclei in Percoll gradients (7) if required.

Protocol 6. Isolation of sea-urchin spermatid nuclei

Equipment and reagents

- Glass homogenizer with a loose-fitting plunger
- Muslin
- 0.34 M sucrose–buffer A (*Protocol 1*)
- ASW (*Protocol 4*) containing 10 mM Hepes, pH 8
- Triton X-100

A. *Identification, removal, and storage of gonads*

1. Remove the urchin shell around the mouth with a pair of scissors to reveal the gonads. Place a small sample of the contents of the gonads onto a microscope slide and view under phase-contrast microscopy (10 $\times$ magnification). For early urchins there are three possibilities:
   (a) The animals are very immature. The cells from both sexes (pre-oocytes and pre-spermatids) look very similar. Try again with another batch of urchins in 1–2 weeks.
   (b) Immature females. Oocytes are much larger than spermatids, but large cells can also be present in very immature males so the identification is ambiguous.
   (c) Immature males. These are unambiguously identified when some motile mature sperm are also present, and are suitable for use.
2. Tease out the (five) gonads intact from the shells if possible, by scraping gently at the connecting tissue with a scalpel blade.
3. Wash the gonads in ASW (*Protocol 4*) containing 10 mM Hepes (pH 8.0). If they are not to be used immediately for isolation of nuclei, put them into 50 ml stoppered plastic tubes, fill up with glycerol, and store at – 80 °C.

B. *Isolation of nuclei*

Carry out all operations at 0–4 °C.

1. Put the gonads from one urchin into a glass homogenizer with ~ 50 ml of 0.34 M sucrose-buffer A (*Protocol 1*). Homogenize with 10 up-and-down strokes of a loose-fitting, hand-held plunger.
2. Filter through eight layers of muslin soaked in 0.34 M sucrose–buffer A and wash the muslin through to give ~ 70 ml homogenate.

3. Centrifuge at 2000 g_{max} for 4 min to pellet sperm nuclei preferentially. The supernatant will be much enriched in spermatid nuclei although there will still be some remaining sperm/sperm heads.
4. Add Triton X-100 to 0.25% (v/v) to the supernatant and leave at 0°C for 3 min. Centrifuge at 3000 g_{max} for 5 min. The pellet contains mainly spermatid nuclei and a few sperm nuclei. Discard the supernatant which contains mainly sperm tails.
5. Wash the spermatid nuclei several times with 0.34 M sucrose–buffer A by repeated centrifugation and resuspension to eliminate any cell debris and sperm tails.
6. Resuspend the pellet of spermatid nuclei in the same buffer at $A_{260} \sim 50$ (read in 1 M NaOH) for digestion with micrococcal nuclease (*Protocol 9*).

2.4 Brain nuclei

The procedure described for the isolation of rat brain nuclei is closely based on a published method (9) for the isolation of nuclear populations heavily enriched in neuronal nuclei (population N^1) or glial nuclei (population N^2) from rabbit cerebral cortex, and cerebellar nuclei, and has also been used for ox brain (10). The cortical nuclei from principal neurones and glia are very different in size and chromatin condensation (as well as nucleosomal repeat length (11)) and this is the basis of their separation by differential centrifugation through sucrose solutions of different densities. The nuclei from cortical neurones are notable for having about half the H1 content (i.e. ~0.5 H1/nucleosome, on average) of the glial nuclei from the same tissue, and of cerebellar nuclei (12).

Protocol 7. Isolation of rat cortical and cerebellar nuclei

Equipment and reagents

- Adult (2–3-month-old) rats
- Motor-driven homogenizer
- Swing-out rotors (e.g. Beckman SW28 and SW50.1) and polyallomer tubes
- 50 mM stock PMSF in propan-2-ol
- 0.34 M sucrose–buffer A (*Protocol 1*)
- 0.32 M, 1.8 M, 2.0 M, 2.2 M, and 2.4 M sucrose solutions each containing 1 mM $MgCl_2$
- Whatman No. 50 hardened filter paper (cat. no. 1450 090)
- Muslin
- Haemacytometer
- Phase-contrast microscope

Method

Carry out all operations at 4°C (in a cold-room). Add PMSF (to 0.25 mM from a 50 mM stock in propan-2-ol) to all solutions immediately before use.

1. Rapidly remove the brains from three adult (2–3-month-old) rats, killed by an approved procedure, into ice-cold 0.32 M sucrose–1 mM $MgCl_2$.

Protocol 7. *Continued*

2. Dissect cortices free of white matter and meninges on Whatman No. 50 hardened filter paper; also dissect the cerebella. Transfer the tissues to ice-cold 2 M sucrose–1 mM $MgCl_2$ and weigh them.
3. Chop the tissue finely with a scalpel and then homogenize in 2 M sucrose–1 mM $MgCl_2$ using a motor-driven homogenizer, to give a 20% (w/v) homogenate.
4. Dilute the homogenate 1:1 with 2 M sucrose–1 mM $MgCl_2$ and filter through two layers of muslin.
5. Collect the nuclei by centrifugation in a swing-out rotor at 90 000 g_{av} for 30 min (e.g. a Beckman SW28 rotor at 25 000 r.p.m.).
6. Resuspend the *cerebellar* nuclei in ~3 ml 0.32 M sucrose and collect by centrifugation at 3000 g_{max}. Repeat this step two or three times. Finally resuspend the nuclei in 0.34 M sucrose–buffer A (*Protocol 1*) and adjust to A_{260} = 20 (read in 1 M NaOH) ready for micrococcal nuclease digestion.
7. Resuspend the *cerebral cortex* nuclei in 3 ml of 2.4 M sucrose–1 mM $MgCl_2$ by gentle vortex-mixing, and transfer to a 5 ml polyallomer ultracentrifuge tube. Overlay the suspension with 1 ml of 2.2 M sucrose–1 mM $MgCl_2$, followed by 1 ml of 1.8 M sucrose–1 mM $MgCl_2$.
8. Centrifuge the step gradient at 130 000 g_{av} for 60 min (e.g. in a Beckman SW50.1 rotor).
   (a) 2.2 M/2.4 M interface: N^1 nuclei (mainly neuronal; large, pale, with a prominent nucleolus when viewed with a phase-contrast microscope);
   (b) pellet: N^2 nuclei (mainly microglial and oligodendroglial; smaller, denser);
   (c) 1.8 M/2.2 M interface: some debris and some N^1 nuclei.
9. Resuspend the N^1 and N^2 nuclei in 0.32 M sucrose–1 mM $MgCl_2$ and wash by centrifugation (3000 g_{max} for 5 min) and resuspension two or three times.
10. Determine the purity of the two populations by counting in a haemacytometer under a phase-contrast microscope. The N^1 population is usually ~ 80% neuronal and the N^2 population 90–95% glial.
11. Resuspend the N^1 and N^2 nuclear pellets in 0.34 M sucrose–buffer A (*Protocol 1*), ready for micrococcal nuclease digestion, and adjust to A_{260} = 20 (read in 1 M NaOH).

3. Isolation and size-fractionation of chromatin

Native soluble chromatin is prepared by the method of Noll *et al.* (13) which avoids the mechanical shearing used in earlier methods, and has been the method of choice since it was introduced in 1975. It involves digestion of chromatin *in situ*, in nuclei, with staphylococcal nuclease (micrococcal nuclease) for a time depending on the average length of chromatin fragments required. The soluble chromatin is size-fractionated by sedimentation through sucrose gradients.

3.1 Preparation of native chromatin

The concentrations of micrococcal nuclease and the digestion times specified below give an idea of the conditions needed for chromatin isolation from rat liver nuclei and chicken erythrocyte nuclei. They are similar for nuclei from other sources digested at the same nuclear concentration. However, the only guaranteed way of ensuring the right conditions to give chromatin of a particular length for nuclei from any particular source is to carry out a time-course of digestion on an analytical scale (1–2 A_{260} units is sufficient) before committing the bulk. To do this, extract the DNA from aliquots taken after various digestion times and analyse in an agarose gel (e.g. 1%) with size markers (*Protocol 10*).

After digestion, nuclei may be lysed at either low (e.g. 0.2 mM Na_2EDTA, pH 7) or high (e.g. 80 mM NaCl) ionic strength to give chromatin in the 10 nm (nucleosome filament) or 30 nm ('solenoid') form, respectively. Hypotonic lysis gives a much better yield of soluble chromatin, which appears to be readily refoldable to the 30 nm form by the addition of NaCl $\geq \sim 70$ mM. One advantage of low ionic strength is that it minimizes the risk of exchange of H1 between sites (14, 15), although it is unclear whether, if exchange of H1 occurs, it perturbs the original distribution.

If it is necessary to determine the repeat length of the chromatin, the nuclei are digested such that the extracted DNA gives a ladder of bands in a 1% or 1.25% agarose gel (*Figure 1*), or 2.5% agarose–0.5% acrylamide composite gel, extending from mononucleosomal to hexanucleosomal DNA and beyond. Since the mononucleosome is invariably 'trimmed' to some degree by the exonuclease activity of micrococcal nuclease, the repeat is best calculated from the difference in the size of successive bands, determined graphically, using appropriate molecular weight markers (11), rather than from the size of mononucleosomal DNA. The repeat lengths for rat liver, chicken erythrocytes, and sea-urchin sperm thus determined are ~200, 215, and 240 bp, respectively.

To prepare 146 bp nucleosome core particles (which do not contain H1) on a small scale, long native chromatin is depleted of linker histones by treatment with NaCl (*Protocol 12B*) and digested with micrococcal nuclease, under conditions determined in preliminary trials, until the DNA is 146 bp

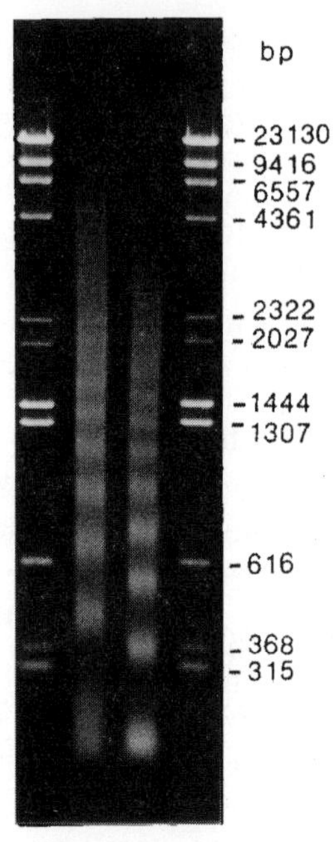

Figure 1. Different nucleosome repeat lengths in sea-urchin sperm nuclei and rat liver nuclei. 1.25% agarose gel electrophoresis (40 mM Tris, 24 mM acetic acid, 2 mM Na_2EDTA, pH 8) of DNA from micrococcal nuclease-digested nuclei from sea-urchin sperm (SUS) and rat liver (RL). The gel (and running buffer) contained 0.5 μg/ml ethidium bromide and was photographed, with UV transillumination, through a red filter using Kodak TriX film. M, marker containing a mixture of *Hin*dIII restriction fragments of bacteriophage λ DNA and *Taq*I fragments of pBR322; sizes are given in base pairs. (The measured repeat lengths are ~ 193 ± 5 and 240 ± 5 bp for RL and SUS, respectively.) (From ref. 5, with permission.)

(with a trace of dimer). They can be purified in sucrose gradients (*Protocol 9*). On a larger scale, core particles are conveniently prepared from chromatin stripped of H1 by gel filtration at the appropriate NaCl concentration (16), as also described elsewhere (17).

Protocol 8. Preparation of native chromatin by micrococcal nuclease digestion

Equipment and reagents

- Micrococcal nuclease (Sigma) (or Boehringer or Worthington). Dissolve at 15 000–17 000 U/ml in 5 mM sodium phosphate, pH 7.0, 0.025 mM $CaCl_2$. Store frozen in aliquots. (1 unit (U) of micrococcal nuclease releases 1 A_{260} unit of oligonucleotide per min at 37 °C at pH 8.8.)
- Centrifuge and Corex glass centrifuge tubes

Method

The following refers to rat liver or chicken erythrocyte nuclei at $A_{260} = 50$ in either 0.34 M sucrose–buffer A (*Protocol 1*) or TMS (*Protocol 3*).

1. Incubate nuclei at 37 °C in a Corex glass centrifuge tube; check with a thermometer that the suspension has reached 37 °C.
2. Add $CaCl_2$ to a final concentration of 1 mM from a 0.1 M stock and mix once quickly.
3. Add the required amount of micrococcal nuclease (e.g. 2 μl/ml nuclei for nuclei at $A_{260} = 50$), mix the enzyme into the suspension well, and incubate at 37 °C for the required time, which will be *approximately*:
   - for short chromatin (1–6 nucleosomes), 6–8 min;
   - for medium length (10–15 nucleosomes; DNA ~ 2–3 kb), 3 min;
   - for long chromatin (> ~ 30 nucleosomes; DNA > ~ 6 kb), 1 min.

   N.B. Swirl the suspension gently from time to time during digestion in order to prevent the nuclei settling; this is particularly important for the longer digestion times. (The enzyme concentration may be doubled (for example), and the time of digestion correspondingly halved to give similar digestion.)
4. Stop the digestion by adding Na_2EDTA, pH 7, to 10 mM (from a 0.1 M stock), and chilling on ice.
5. Collect the nuclei by centrifugation at 2000 g_{max} for 4 min at 4 °C (e.g. in a Sorvall SS-34 rotor). Discard the supernatant and drain the pellet well by inverting the tube on paper tissue.
6. Lyse the nuclei in 0.2 mM Na_2EDTA pH 7 (e.g. in a volume equal to the original volume of nuclei digested) by gently pipetting up and down with a Pasteur pipette. If the DNA is very long just swirl gently, but leave on ice for at least 30 min to ensure maximum lysis (or dialyse overnight against 0.2 mM Na_2EDTA, pH 7). The resulting solution will be opalescent but not viscous. The nuclei can be lysed in as little as ~ 1/3 of the volume of the original nuclear suspension; the yield of chromatin falls as the volume is reduced further.
7. Centrifuge long chromatin at 2000 g_{max} and short chromatin at 3000 g_{max} for 4 min (e.g. in a Sorvall SS-34 rotor) and carefully decant the supernatant (soluble chromatin) without disturbing the pellet (nuclear debris). Short chromatin gives a clear solution; long chromatin is faintly opalescent.
8. Measure the absorbance (e.g. of 10 μl) at 260 nm in water. 1 ml nuclei (at $A_{260} = 50$, read in 1 M NaOH) usually gives 1 ml long chromatin at A_{260} ~ 15–25 (read in water) or short chromatin at A_{260} ~ 35.

   NB As a rule of thumb, $A_{260} = 10$ corresponds to ~ 1 mg/ml chromatin or ~ 0.5 mg/ml each of histones and DNA.

Protocol 8. *Continued*

9. Check the DNA size and the integrity of the histone complement by gel electrophoresis (*Protocols 10 and 11*, respectively).

3.2 Size-fractionation of chromatin

The best method by far for size-fractionation of nucleosomes and oligonucleosomes is by sedimentation through sucrose gradients. Gel filtration has low

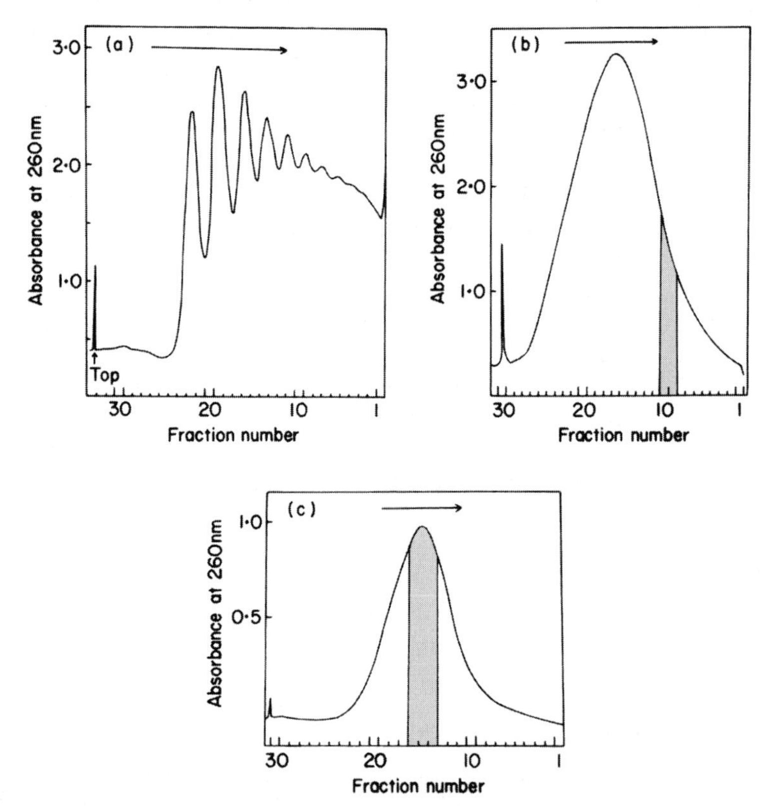

Figure 2. Sucrose-gradient fractionation of the soluble chromatin from rat liver nuclei digested for different times with micrococcal nuclease. (a) Short oligomers; in this case rat liver nuclei at A_{260} = 50 were digested for 3 min at 37 °C with 2.5 μl (37.5 U) of Worthington micrococcal nuclease/ml nuclei. The chromatin was fractionated in 5–30% linear sucrose gradients (see text) which were centrifuged for 20 h at 27 000 r.p.m. in a Beckman SW27 rotor at 4 °C. (b) Long oligomers; in this case digestion was for 1 min with 1 μl (15 U) enzyme/ml nuclei and centrifugation was at 15 500 r.p.m. for 15 h, otherwise as in (a). The stippled region indicates fractions taken for removal of H1; see (c). (c) Sedimentation profile of fractions from the stippled region in (b) dialysed, concentrated, treated with 0.55 M NaCl, and centrifuged though a 5–30% sucrose gradient containing 0.55 M NaCl in an SW27 rotor for 16 h at 18 000 r.p.m. The DNA size of the three pooled peak fractions (not shown) indicated a weight-average size of 77.4 nucleosomes for the chromatin (weight-average DNA size ~ 15.5 kb). (From ref. 19, with permission.)

resolution and there may also be a risk of slight loss of H1; e.g. gel filtration of long chicken erythrocyte chromatin on Bio-Gel A-150m in 60 mM NaCl was sufficient to affect the folding of the chromatin (18). Resolution in sucrose gradients is better at low ionic strength (e.g. 10 mM Tris–HCl, pH 7.5, 0.1 mM Na_2EDTA, 0.25 mM PMSF) than at high ionic strength.

We routinely use simple 5–30% (w/v) linear gradients (19), rather than isokinetic gradients (20). Although gradients can, of course, be poured manually one at a time, we usually pour six gradients simultaneously using a six-way stream splitter between the gradient reservoirs and a six-channel pump (e.g. Desaga STA peristaltic). Gradients may be poured at room temperature and should then be kept at 4 °C for up to a few hours until needed.

Fractionation after centrifugation can be carried out in various ways. We pump from the bottom (through a 10 cm syringe needle (i.d. ~1 mm) clamped so that it can be lowered to within a millimetre of the bottom of the gradient tube) through a UV (260 nm or 280 nm) flow cell connected to a chart recorder and into a fraction collector. (For alternative methods of fractionation see ref. 20.)

For short chromatin, the absorbance profile will show distinct peaks corresponding to mononucleosomes, dinucleosomes, etc. (the resolution becoming steadily worse for the higher oligomers) (*Figure 2a*), and the size fractions of interest can be identified and taken directly. Their DNA contents are shown in *Figure 3*. For medium/long chromatin the profile will show a broad distribution, without distinct peaks (*Figure 2b*), and fractions of the required oligonucleosome length must be identified from their DNA contents.

Protocol 9. Sucrose-gradient fractionation of micrococcal nuclease digested chromatin

Equipment and reagents

- Equipment for pouring linear sucrose gradients: pump, gradient-forming chamber, magnetic stirrer
- Equipment for fractionating sucrose gradients: pump, UV monitor (flow through), chart recorder, fraction collector
- Swing-out rotor and tubes (e.g. Beckman SW28 rotor with 17 ml or 38 ml polyallomer tubes depending on the size of gradient required)
- Sucrose solutions: 5% and 30% (w/v) in 10 mM triethanolamine–HCl (or Tris–HCl), pH 7.5, 1 mM Na_2EDTA, and 0.5 mM PMSF
- Dialysis tubing (Medicell). Boil appropriate lengths for 20 min in 2 mM Na_2EDTA, pH 7, rinse 5 or 6 times with distilled H_2O, and store in 10% ethanol at 4 °C.

Method

1. Pour sucrose gradients of the appropriate size and number to accomodate the volume of chromatin solution to be fractionated. Calculate the volume of the gradient so that the gradient plus the chromatin solution fills the tube to within ~2 mm of the brim. A maximum volume of 3 ml and about 200 A_{260} units is recommended for 38 ml tubes (~35 ml

Protocol 9. *Continued*

gradients) in a Beckman SW28 rotor; and a maximum of 1.5 ml and about 100 A_{260} units for 15.5 ml gradients in 17 ml tubes. Lower loadings give better resolution.

2. Layer the soluble chromatin solution carefully on to the gradients using a Pasteur pipette. Taking care not to disturb the gradients, screw on the bucket caps and load the buckets carefully onto the rotor. Centrifuge at 4°C for the appropriate time and at the appropriate speed, chosen according to the length of the chromatin (see below). Set the centrifuge to switch off without the brake, to avoid disturbing the gradients. Suitable centrifugation conditions for overnight (16 h) runs, which may often be convenient, are as follows:
   (a) for short chromatin (e.g. good resolution of mono- to hexanucleosomes): 104 000 g_{av} (e.g. 28 000 r.p.m. in a Beckman SW28 rotor);
   (b) for medium/long chromatin: 37 000 g_{av} (e.g. 16 000 r.p.m. in a Beckman SW28 rotor).

   If necessary, adjust the speeds and times according to the relationship:

$$\frac{t_1}{t_2} = \frac{v_2^2}{v_1^2}$$

   where the speeds (in r.p.m.) and times given above are t_1 and v_1. For more rapid analysis use rotors appropriate for higher speeds (e.g. a Beckman SW40 rotor or SW50 rotor), but remember that the tubes have a smaller capacity than those for the SW28 (e.g. 12 ml for SW40 and 5 ml for SW50, compared with 17 ml or 38 ml for the SW28 rotor) which limits the amount of chromatin that may be loaded in one run.
3. Fractionate the gradients by pumping them out from the bottom through a UV monitor at 260 nm or 280 nm to monitor the absorbance continuously, and deliver into a fraction collector (e.g. set to collect 1 ml fractions). Remember to take account of the delay between the flow cell and the fraction collector when you come to identify fractions based on the absorbance profile.
4. Determine the DNA size (*Protocol 10*) by analysing the sucrose-gradient fractions either directly or after dialysis into a suitable buffer, e.g. 10 mM triethanolamine–HCl, pH 7, 0.2 mM Na_2EDTA (the latter to ensure inhibition of any residual micrococcal nuclease that might have remained attached during the fractionation).

3.3 Analysis of histone and DNA content

Short DNA fragments (from mono- to penta- or hexanucleosomes) are more conveniently analysed in 1% or 1.25% agarose gels (e.g. *Figure 1*) than in composite 2.5% polyacrylamide/0.5% agarose gels (e.g. *Figure 3*), although

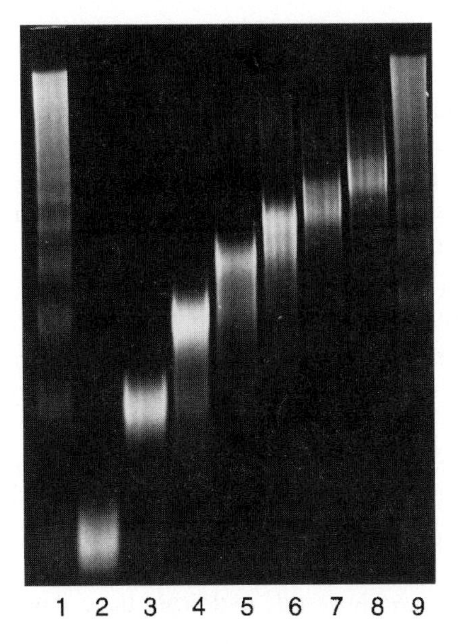

Figure 3. Determination of the nucleosome content of the chromatin fragments by measurement of their DNA sizes. DNA content of mono- and small oligonucleosome peaks from the sucrose gradient in *Figure 2a*, analysed in a 2.5% polyacrylamide/0.5% agarose gel in Tris–acetate buffer (56), stained with ethidium bromide, and then photographed as in *Figure 1*. Lanes 2–8, mono-, di- ... up to nominal heptanucleosome peaks. Lanes 1 and 9, DNA from an unfractionated micrococcal nuclease digest. (From ref. 19, with permission.)

both give good resolution of the DNA from successive nucleosome oligomers. For longer DNA fragments 0.7% or 0.8% agarose gels are suitable.

Protocol 10. Determination of the DNA content of chromatin fractions

Equipment and reagents

- Horizontal (submarine) DNA gel-electrophoresis apparatus and power pack
- UV transilluminator
- Agarose
- 10 mg/ml ethidium bromide in water, stored in a foil-covered bottle. **Caution**: ethidium bromide is a mutagen
- 1% (w/v) bromophenol blue in methanol
- *either* TAE: 40 mM Tris base, 24 mM acetic acid, 2 mM EDTA, pH 8 *or* TBE: 89 mM Tris base, 89 mM boric acid, 2 mM EDTA, pH ~8.3
- 10% SDS
- 5M NaCl
- Chloroform:isoamyl alcohol (24:1 (v/v))

Method

1. Take a chromatin sample (~0.04 A_{260} unit; ~2 μg DNA) from a sucrose-gradient fraction, either directly or after dialysis (see above), and dilute it to (e.g.) 350 μl with water in a 1 ml microcentrifuge tube. At room temperature add 10% SDS (1/7th vol), mix well, and add 5 M

Protocol 10. *Continued*

NaCl (2/7th vol). Add an equal volume of chloroform/isoamyl alcohol (24:1 (v/v)) and vortex to give an emulsion. Centrifuge for 5 min at room temperature in a microcentrifuge. Remove the (top) aqueous layer carefully into a clean tube and re-extract. To the aqueous layer add 5 vol. chilled absolute ethanol and chill in ethanol/dry ice for 20 min or overnight at – 20 °C to precipitate the DNA.

2. Collect the DNA by centrifugation in a microcentrifuge at full speed for 15 min, wash the pellet with 70% ethanol and dry under vacuum. (If the DNA is extracted directly from the sucrose gradient the pellet might contain sucrose; redissolve in water and reprecipitate with ethanol.)
3. Dissolve the DNA in (e.g.) 20 μl of TAE buffer (or TBE) containing 20% (w/v) sucrose, 1 μg/ml ethidium bromide, add 2 μl bromophenol blue and load 10 μl onto a ~ 12 cm 1% or 1.25% agarose gel (short to medium chromatin) or 0.7% gel (long chromatin) in TAE or TBE buffer containing ethidium bromide (0.5 μg/ml) in both the gel and the running buffer. Carry out electrophoresis at ~ 5 V/cm for ~ 2.5–3 h (1% or 1.25% gel) or at ~ 1 V/cm for ~ 12 h (0.7% gel).
4. Visualize the DNA by transillumination on a UV light box at 305 nm. If required, photograph the gel through a red filter using Kodak Tri-X 35 mm film, or similar.

Analysis of the histones in SDS gels is necessary to assess the quality of the chromatin, in particular: (1) to determine whether there has been any proteolysis (H1/H5 (21) and then H3 (13) are the most susceptible (22)); and (2) to assess the linker histone (H1, H5) content. The linker histone content will be lower for short oligonucleosomes (up to 6-mer (14)) than for long chromatin, due to trimming of the terminal linkers which stabilize H1(H5)-binding to the terminal nucleosomes. Less than the canonical ~ 1 H1 per nucleosome (or 0.9 H5 + 0.4 H1 for chicken erythrocyte chromatin (23)) is easily detected by densitometry from the staining of the bands relative to the staining of H4, which is stably bound and also not easily proteolysed (24). The gel should also contain a sample of long chromatin or of whole nuclei boiled up, or of total acid-extracted histones, as a 'standard'.

We use a modification (24, 25) of the original method of discontinuous SDS/polyacrylamide gel electrophoresis (26), to obtain better resolution of the core histones. The acrylamide:*bis*acrylamide ratio in the separating gel is lower, the concentration of the Tris–HCl buffer in the resolving gel is higher, and the concentration of the Tris/glycine/SDS running buffer is also altered. *Figure 4* shows a 12 cm SDS/18% polyacrylamide gel of rat liver, chicken erythrocyte, and sea-urchin sperm histones (5). The resolution can be further

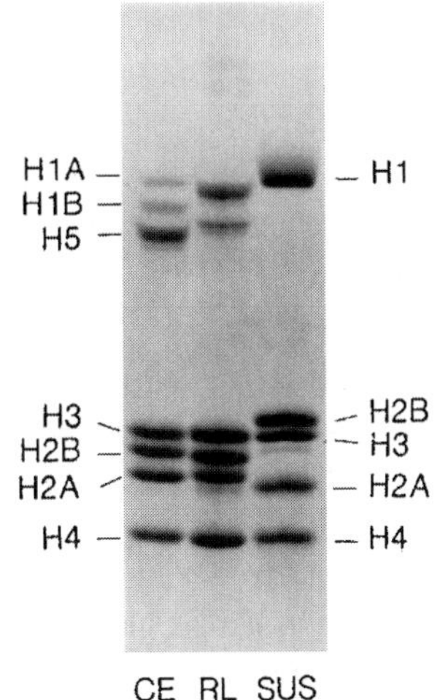

Figure 4. Histone contents of nuclei from sea-urchin sperm (*Echinus esculentus*) (SUS), chicken erythrocytes (CE), and rat liver (RL) determined by SDS/18% polyacrylamide gel electrophoresis. The gel was stained with Coomassie Brilliant Blue R-250. (Note the presence of distinct H1 variants in CE and SUS, compared with RL (which is typical of many mammalian tissues), and in the case of SUS also a sperm-specific H2B.) (From ref. 5, with permission.)

improved with a longer gel (24). Note that the two H1 bands in chicken erythrocytes contain several (six) subtypes.

Protocol 11. Analysis of the histone content of chromatin fractions

Equipment and reagents

- Vertical polyacrylamide gel electrophoresis apparatus
- An SDS/polyacrylamide gel (e.g. ~6 cm long, 1 mm thick) with an 18% polyacrylamide separating gel (acrylamide:bisacrylamide, 30%:0.15% (w/w)) containing 0.75 M Tris–HCl, pH 8.8, and a 3% stacking gel (acrylamide:bisacrylamide, 10%:0.5% (w/w)) containing 0.12 M Tris–HCl, pH 6.8)
- SDS-gel running buffer: 1 litre of 5 × buffer contains 30 g Tris base, 140 g glycine, 50 ml 10% (w/v) SDS; pH after dilution is 8.4
- SDS-gel sample buffer: 6.9 ml water, 1.0 ml glycerol, 0.1 ml 2-mercaptoethanol, 1 ml SDS (10% (w/v)), 1.0 ml 0.5 M Tris–HCl, pH 6.8, 5–10 μl 1% bromophenol blue. (Store frozen in aliquots.)
- 50% (w/v) trichloroacetic acid (TCA; AR grade)
- Acetone (AR grade)/10 mM HCl
- Coomassie Brilliant Blue R-250 (0.1% w/v in methanol/acetic acid/H_2O (5:1:5 v/v)

Method

1. Take chromatin samples from the sucrose-gradient fractions equivalent to ~0.2 A_{260} units, which will contain ~10 μg (total) of histones, into a microcentrifuge tube and add an equal volume of water.

Protocol 11. *Continued*

2. Add an equal volume of 50% TCA, mix and leave on ice for 15 min to precipitate proteins and DNA.
3. Collect the precipitate by centrifugation for 5 min at full speed in a microcentrifuge, wash the pellet with acetone/10 mM HCl, recentrifuge, wash with acetone and dry in air or under vacuum. If the pellet after TCA precipitation is sticky (sucrose), redissolve in water and repeat the precipitation step.
4. Dissolve the pellets in 20 μl SDS-gel sample buffer (already incubated in a boiling water bath) in microcentrifuge tubes, stopper the tubes, pierce a hole in each lid with a syringe needle, and heat at 100°C for 1 min.
5. Cool, centrifuge briefly (this will also pellet very long DNA), and apply the samples to an SDS/18% polyacrylamide gel. (Ideally, run in parallel a sample of total acid-extracted histones (0.2 M H_2SO_4 extraction followed by precipitation with 25% TCA and washing as in step 3 above) or of whole nuclei boiled up in SDS-gel sample buffer. Comparison of the test sample with this after staining (by densitometry if necessary) will reveal any histone proteolysis or loss.) Run the gel at 30 mA until the dye reaches the bottom of the gel (~ 2 hours). For better resolution of the core histones, particularly H3 and H2B, use a longer (e.g. ~ 12 cm) gel, especially if densitometry is intended.
6. Fix the gel in methanol/acetic acid/H_2O (5:1:5 (v/v)) and stain with Coomassie Brilliant Blue R-250 (0.1% (w/v) in fix solution), both for 15–30 min at room temperature. Destain with 5% methanol/7.5% acetic acid in H_2O with gentle agitation, either at room temperature or with warming for faster destaining, and with the inclusion of foam rubber pieces to adsorb the free stain.

4. Removal of linker histones from native chromatin

Linker histones may be removed conveniently from micrococcal nuclease digested chromatin fragments of any length by displacement with the appropriate concentration of NaCl (0.5–0.65 M; see below) and sedimentation through a sucrose gradient containing the same salt concentration at 4°C. The nucleosome repeat length of the chromatin is unaffected by this procedure, as shown by comparing the micrococcal nuclease digestion patterns before and after linker-histone depletion. Stripping of linker histones may also be carried out (27) at much lower NaCl concentrations (80 mM NaCl) by passage of the chromatin twice over a DNA-cellulose column which competes for H1(H5) which can exchange under these conditions (14, 15). For stripping the sperm-specific H1 from sea-urchin sperm chromatin it is the preferred method, because stripping by the sucrose-gradient method requires NaCl concentrations >0.65 M, with the accompanying risk of nucleosome sliding.

Slightly different procedures are appropriate/convenient for the generation of defined, short H1-stripped fragments (e.g. mono-, di-, or trinucleosomes, etc.) on the one hand, and medium or long H1-stripped oligomers, by the sucrose-gradient method:

(a) In the case of short chromatin fragments, these are isolated by size-fractionation (*Protocol 9*) and then recentrifuged through a sucrose gradient containing NaCl (e.g. 0.5 M for removal of H1 from rat liver chromatin). This procedure is adopted, rather than fractionating the mixture of oligomers produced by micrococcal nuclease digestion directly in a 0.5 M NaCl-containing gradient, because the resolution of the different oligomers under these conditions is poor.

(b) In the case of medium/long chromatin fragments, which are poorly resolved in sucrose gradients anyway (see section 3.2 and *Figure 2b*), nothing is lost by adopting the faster procedure of applying the digest directly to NaCl-containing gradients and then identifying the fractions of the required size by analysis of the DNA as for unstripped chromatin (*Protocol 10*). Medium or long oligomers which have already been size-fractionated may, of course, also be H1-depleted in a separate centrifugation step in a gradient containing NaCl (*Figure 2c*), as for short oligomers.

Protocol 12. Salt-stripping of linker histones in sucrose gradients

Equipment and reagents

- As for *Protocol 9* except that the sucrose solutions contain 0.5 M or 0.55 M NaCl
- SDS/18% polyacrylamide gel (*Protocol 11*)

A. *Short fractionated oligomers*

1. Prepare soluble chromatin of the required size range (*Protocol 8*), fractionate to give mononucleosomes/short oligonucleosomes of the required size (*Protocol 9*), and dialyse against 10 mM triethanolamine-HCL, pH 7, 0.1 mM EDTA, 0.25 mM PMSF.
2. Concentrate in the dialysis bag with solid sucrose at 4°C, and redialyse to give a ~2-fold concentration relative to the sucrose-gradient fractions.
3. Add NaCl from a 2 M stock to 0.5 M (rat liver chromatin) or 0.55 M (chicken erythrocyte chromatin) in a steady stream with constant swirling (there may be transient opalescence), and layer onto prepared sucrose gradients of the same composition as in *Protocol 9* but also containing 0.5 M or 0.55 M NaCl, respectively. The volumes and

Protocol 12. *Continued*

amounts loaded should be no greater than those specified in *Protocol 9*. (Note that 0.55 M NaCl is sufficient to strip short chicken erythrocyte oligomers, whereas 0.65 M is required for long chromatin (see part B)).

4. Centrifuge the gradients, bearing in mind when selecting the speed (see *Protocol 9*) that H1,H5-depleted chromatin fragments sediment about half as fast as fragments of the same size but unstripped, because the chromatin is now in its extended form and will sediment more slowly.
5. Fractionate the gradients (*Protocol 9*), pool the required fractions and dialyse against 10 mM triethanolamine–HCl (*or* 10 mM Tris–HCl), pH 7, 0.1 mM Na_2EDTA, 0.25 mM PMSF.
6. TCA-precipitate a small sample (~0.2 A_{260} unit) and analyse in an SDS/18% polyacrylamide gel to check for complete removal of the linker histones (*Protocol 11*).

B. *Medium/long chromatin*

This may be treated exactly as in part A, with two separate centrifugation steps of the appropriate speed. Alternatively, use a shorter procedure, as discussed above.

1. Prepare soluble chromatin of the required size range (*Protocol 8*).
2. Add NaCl to 0.5 M (for the removal of H1 from rat liver chromatin) or 0.65 M (removal of H1,H5 from chicken erythrocyte chromatin) and proceed as in part A (steps 3, 4).
3. Identify fractions of the required length, *either* by removing samples from a series of fractions (in the range estimated to be of interest) and determining the DNA size (*Protocol 10*), meanwhile keeping the fractions at 4°C, and then pooling and dialysing as above; *or* by dialysing individual fractions as above, and then determining the DNA size before pooling on this basis.
4. TCA-precipitate a small sample (~0.2 A_{260} unit) and analyse in an SDS/18% polyacrylamide gel to check for complete removal of the linker histones (*Protocol 11*).

5. Isolation of linker histones

Linker histones (H1 and its variants and subtypes) may be isolated directly from nuclei either by salt extraction with NaCl in a range from 0.5 to 0.65 M depending on the variant, or by extraction with 5% perchloric acid. The latter is convenient for certain purposes (e.g. if the H1 is required only for analysis in SDS/polyacrylamide gels, or for protein-chemical studies, amino-acid analysis, mass spectrometry, etc.) and it also minimizes proteolysis. However, as sometimes practised (e.g. prolonged exposure to perchloric acid followed by acetone precipitation or trichloroacetic acid precipitation), there is a distinct risk of irreversible denaturation (e.g. even overnight exposure to pH 2 (HCl) at 4 °C resulted in altered properties compared with salt-extracted H1 (28)), and should be avoided if the H1 is to be used for physical studies or for binding to DNA or chromatin. On the other hand, *brief* (10–15 min) perchloric acid extraction at 4 °C, followed by immediate neutralization and dialysis against phosphate buffer, which promotes folding of the globular domain (29), gives H1 with DNA- and chromatin-binding properties that appear to be indistinguishable from those of salt-extracted H1. Nonetheless, salt extraction should be used whenever convenient, and provided that it is not accompanied by problems of proteolysis.

We therefore use salt extraction of rat liver H1 and of chicken erythrocyte H1 and H5 (30); however, for sea-urchin sperm H1 the mild perchloric-acid extraction procedure (31) is preferred, because appreciable proteolysis can occur during salt extraction in this case, even in the presence of the common protease inhibitors, probably due to the release of proteolytic enzymes from the acrosome in the sperm head. Spermatid H1 is also conveniently extracted by the mild perchloric-acid procedure. Whether salt- or acid-extracted, the linker histones may be purified by cation-exchange chromatography on carboxymethyl cellulose. They may be further purified (e.g. for mass spectrometry) by reversed-phase chromatography using acetonitrile/trifluoroacetic acid (*Protocol 15*), but it would be very unwise to use material purified in this way in binding studies without thorough scrutiny of its behaviour, and certainly not without prior dialysis against phosphate buffer, pH 7, to stabilize the folded structure.

Protocol 13. Isolation of linker histones by salt extraction

Equipment and reagents

- Gradient former (e.g. BioRad model 395, cat. no. 165-2001) and peristaltic pump
- Fraction collector
- Screw-capped plastic Falcon tubes (or similar)
- Corex glass centrifuge tubes
- Carboxymethyl cellulose (Whatman CM52) column packed in a 10 ml syringe (1 ml bed per 5 mg total protein) and equilibrated with 10 mM Na phosphate, pH 7 at 4 °C
- 0.35 M NaCl wash buffer: 10 mM Tris–HCl, pH 7, 0.5 mM PMSF, 0.35 M NaCl
- 0.65 M NaCl extraction buffer: 10 mM Tris–HCl, pH 7, 0.5 mM PMSF, 0.65 M NaCl
- Buffers for column chromatography: 10 mM Na phosphate, pH 7 (column buffer); column buffer containing 0.35 M NaCl; column buffer containing the relevant NaCl concentrations (e.g. 0.3 M and 0.8 M) for the required gradient (see below)
- SDS/18% polyacrylamide gel (*Protocol 11*)

Method

The procedure described is for *chicken erythrocyte linker histones* on a fairly large scale; it can clearly be scaled down. The procedure for *rat liver H1*, which would be carried out conveniently on about a tenfold smaller

Protocol 13. *Continued*

scale (with respect to starting quantities of nuclei), is very similar, except that step 2 is omitted because the nuclei lyse, and 0.3 M NaCl is replaced by 0.25 M NaCl in steps 8 and 9. Rat liver H1 elutes as a single peak at ~0.35 M NaCl in step 9.

Carry out all operations at 0–4 °C.

1. For convenience, carry out the extraction in screw-capped plastic Falcon tubes (or similar). Add PMSF to 1 mM to the nuclei to be extracted. Collect the nuclei (e.g. 20 000 A_{260} units for a large-scale preparation) by centrifugation (2000 g_{max} for 5 min).
2. Wash with 200 ml 0.35 M NaCl wash buffer and collect the nuclear pellet by centrifugation. (The HMG proteins are removed in this step.)
3. Add 200 ml of 0.65 M NaCl extraction buffer; the nuclei lyse to give a jelly-like mass of chromatin. Leave the tubes on a rocking platform or end-over-end mixer for 12 h in the cold room which minimizes shearing of the chromatin during the H1/H5 extraction.
4. Centrifuge at ~102 000 g_{av} (e.g. at 27 000 r.p.m. in a Beckman SW28 rotor) for 8 h at 4 °C, or faster in a lower-capacity, higher-speed rotor for a smaller-scale preparation, and discard the compressed rubbery pellet. (In principle, this could be used to extract core histones (with 2 M NaCl), but the pellet is difficult to manipulate, and is best discarded; alternative procedures may be used for the isolation of core histones (refs 1, 32; see also Chapter 2).
5. Dialyse the clear supernatant against several changes of 10 mM Na phosphate, pH 7.0, 0.25 mM PMSF at 4 °C.
6. Transfer the dialysate to Corex glass centrifuge tubes and centrifuge at 10 000 g_{max} for 15 min (e.g. in a Sorvall SS-34 rotor) to remove any precipitate. (A small amount of precipitate may form due to rebinding of the linker histones to small, soluble chromatin fragments arising from shearing, or even to RNA. It reduces the yield of linker histones somewhat, particularly of H5 relative to H1 in the case of chicken erythrocyte histones, since H5 binds before H1 as the ionic strength is lowered during dialysis.)
7. Discard the pellet and check the A_{230} of the supernatant. There is usually ~70 mg of H1/H5 at this stage (taking A_{230} = 1.85 for 1 mg/ml).
8. Pump the clear supernatant onto a CM52 column (30 ml) equilibrated in 10 mM Na phosphate, pH 7, in the cold-room. Wash the column through with 3–4 column vol. of column buffer (10 mM Na phosphate, pH 7) and then with the same volume of column buffer containing 0.3 M NaCl.
9. Elute with a gradient (~12 column vol.) of 0.3–0.8 M NaCl in column

buffer. Collect 1 ml fractions and monitor the absorbance at 230 nm (histones have a low tyrosine content and no tryptophan, hence low absorbance at 280 nm). Chicken erythrocyte linker histones elute as two peaks corresponding to H1 (midpoint of elution ~0.4 M NaCl) and H5 (~0.6 M NaCl).

10. Analyse alternate fractions across the part of the profile with 230 nm absorbance by electrophoresis in an SDS/18% polyacrylamide gel (*Protocol 11*). Precipitate aliquots corresponding to 0.004 A_{230} unit with an equal volume of 50% (w/v) trichloroacetic acid and wash, etc. (*Protocol 11*) before dissolving in SDS-gel loading buffer. Alternatively, if the fractions are concentrated enough, samples containing 0.004 A_{230} unit may be taken directly into SDS-gel loading buffer (e.g. into an equal vol. of 2 × buffer, provided the final sample volume is not greater than ~20 μl). (Note that loading equal amounts of protein across the peaks (rather than equal volumes of fractions) indicates clearly the purity of the material at the leading and trailing edges.)

11. Pool the fractions containing pure H1 or H5 and *either* dialyse against 10 mM Na phosphate, pH 7, and store frozen, *or* dialyse against water or 0.1% (w/v) fresh ammonium bicarbonate and lyophilize for storage.

12. Determine the concentrations of H1 or H5 stock solutions by amino-acid analysis, with reference to the known amino-acid sequences (33). The concentrations may also be estimated fairly accurately from the A_{230} (A_{230} = 1.85 for 1 mg/ml (34)). This scale of preparation should give ~20 mg pure H1 and 30 mg pure H5 from chicken erythrocyte nuclei.

Protocol 14. Isolation of linker histones by a rapid 5% perchloric-acid extraction/neutralization procedure

Equipment and reagents

- Ion-exchange chromatography equipment/ reagents as in *Protocol 13*
- 5% perchloric acid, diluted from a 60% (v/v) stock solution
- 1 M triethanolamine
- 10 mM Na phosphate, pH 7
- PMSF (50 mM in propan-2-ol)

Method

The method is described for the large-scale preparation of chicken erythrocyte H1/H5, for direct comparison of yields with the salt-extraction method (*Protocol 13*), and is easily scaled down. (However (see above), in this laboratory erythrocyte linker histones are usually salt-extracted.) The procedure is also applicable to nuclei from other souces (e.g. rat liver, sea-urchin sperm).

Protocol 14. *Continued*

1. Divide 20 000 A_{260} units between, say, eight Corex glass centrifuge tubes on ice, collect the nuclei by centrifugation at 3000 g_{max} for 10 min, and lyse by adding 5 ml 0.2 mM Na_2EDTA to each tube.
2. Add 1/11th vol. of ice-cold 60% perchloric acid and mix on ice for 15 min.
3. Centrifuge at 10 000 g_{max} for 10 min at 4°C. Decant the supernatant carefully into a beaker on ice and immediately neutralize with 1 M triethanolamine using a pH meter. Clarify the solution if necessary by centrifugation at 10 000 g_{max} for 10 min.
4. Dialyse at 4°C against 10 mM Na phosphate, pH 7, and purify the linker histone(s) by chromatography on carboxymethyl cellulose (*Protocol 13*). Erythrocyte H1 and H5, and rat liver H1, elute as above (*Protocol 13*, step 9); sea-urchin sperm H1 elutes at 0.65 M NaCl in a gradient of 0. 3–1.0 M NaCl in column buffer, and spermatid H1 (phosphorylated) at ~0.45 M NaCl. .
5. Redialyse and store frozen in aliquots. The yields are higher than with the salt-extraction procedure (*Protocol 13*), giving from the same initial amount of nuclei, 26 mg H1 and 77 mg H5 (in this procedure there is no selective loss of H5; see *Protocol 13*, step 6). If necessary (e.g. for electrospray-ionization mass spectrometry), purify small samples of linker histones by reversed-phase HPLC (*Protocol 15*).

Protocol 15. Purification of linker histones by reversed-phase HPLC

Equipment and reagents

- C-4 Hi-pore reversed-phase column (Bio-Rad; 5 μm, 25 × 0.46 cm)—or similar—with a guard column attached. (A C-8 column is also satisfactory.)
- 0.3% (v/v) aqueous TFA (Sequanal grade; Pierce)
- 0.3% (v/v) TFA in acetonitrile (Fisher, Far UV grade)

Method

1. Dissolve the sample in 0.3% aqueous TFA and apply to the column.
2. Monitor at 230 nm. Elute with a gradient of acetonitrile in 0.3% TFA (e.g. 20–50% acetonitrile in 60 min; the various H1s elute in the range from 30 to 40% acetonitrile) and then wash the column with 100% acetonitrile.
3. Dry-down samples containing 0.004 A_{230} from the fractions likely to be of interest for analysis by SDS/polyacrylamide gel electrophoresis (*Protocol 11*) and pool on this basis.

6. Characterization of the phosphorylation state of linker histones

Linker histones in somatic cells undergo cell-cycle dependent phosphorylations, both at S-phase (one or a few sites phosphorylated, depending on the H1 subtype) and at metaphase (several sites phosphorylated) (35, 36). Most of these phosphorylations are now known to occur at serine or threonine residues in '–SPKK–' or closely related sequences (–S/T–P–X–K/R–). Phosphorylation of linker histones at these sites also occurs in certain cell types at particular stages of differentiation when new variants are synthesized (see ref. 8 for references). For example, H5 in immature nucleated erythrocytes, which replaces most of the H1, is initially multiply phosphorylated and becomes dephosphorylated during erythrocyte maturation, resulting in denser packing of nuclear chromatin and shut-down of transcription. Similarly, during spermatogenesis in echinoderms (e.g. sea-urchins), the cell-type specific H1 variant, spH1, is heavily phosphorylated in spermatids (~ 6–9 sites, depending on the species) and subsequently dephosphorylated, with increased chromatin condensation/packing, in mature sperm. Upon fertilization, the sperm histones become rephosphorylated in the egg cytoplasm and are subsequently replaced by the maternal cleavage-stage histone; this is itself phosphorylated and is, in turn, replaced by somatic H1 during the subsequent cell divisions in the early embryo.

A molecular understanding of the chromatin remodelling that accompanies reversible phosphorylation events requires methods, first, for detecting linker-histone phosphorylation; second, for determining the total number of phosphorylation sites; third, for identifying the positions of the modifications in the amino-acid sequence. Suitable methods are described below. To inhibit phosphatase action, phosphatase inhibitors (e.g. *p*-chloromercuriphenylsulphonate (37)) or β-glycerophosphate should be included during isolation of the phosphorylated histones, e.g. by perchloric-acid extraction (*Protocol 14*) or salt extraction (*Protocol 13*).

6.1 Detection of phosphorylation by gel electrophoresis

SDS/18% polyacrylamide gel electrophoresis (*Protocol 11*) in long (12 cm or longer) gels is often adequate to detect multiple phosphorylation of H1 (e.g. in metaphase *vs.* interphase cells from eukaryotic cell cultures; and in sea-urchin spermatids *vs.* mature sperm) which is reflected in a shift to a species of lower mobility. The shift is abolished by treatment of the sample with alkaline phosphatase before electrophoresis. If ^{32}P-labelling of cells is possible this will confirm that the slower band is phosphorylated.

Better resolution of unphosphorylated and phosphorylated forms and, in some cases, detection of forms with intermediate phosphorylation (see, for example, ref. 35), may be achieved in other types of gel systems, e.g. the well-established acetic acid–urea (38) or Triton–acetic acid–urea systems (39), which give good resolution (see ref. 40). However, in some instances these systems, although very effective, are not particularly convenient because of the long pre-running/running times for the gels (typically, 15% acrylamide) which may be at least 12 cm long, and often twice as long. The more recently described neutral Hepes–histidine–urea gels (41) provide an alternative. At neutral pH the phosphate group carries a charge of between – 1.5 and – 2 at pH 7 (compared with – 1 at pH 4.5, a typical pH for acid–urea gels) and hence the separation of phosphorylated and unphosphorylated forms is increased. For example, the electrophoretic mobilities of the interphase and metaphase (phosphorylated) forms of HeLa H1 differ by about 10% in Hepes–histidine–urea gels, but only by about 5% in the other gel systems mentioned above (41). Gels only 5 cm long, containing 10% acrylamide (acrylamide: *bis*acrylamide, 30%:0.6% (w/w)) in the separating gel and 6% acrylamide in the stacking gel, which run in about 2.5 hours, are convenient.

6.2 Determination of the total number of phosphorylated sites by mass spectrometry

Methods are now available for the determination of the molecular masses of proteins (i.e. as well as peptides) by mass spectrometry (42, 43). The total number of phosphorylated sites may be unambiguously determined by comparison of the molecular masses of the phosphorylated and unphosphorylated (or dephosphorylated, by alkaline phosphatase) forms of the same molecule (e.g. of (phosphorylated) spermatid and (unphosphorylated) sperm H1 (44)) or, if the amino-acid sequence of the protein is known, from the difference between the observed mass and that calculated from the sequence. In each case, the difference in mass divided by 80 (the mass of a phosphate group is 79.8 Da) gives the number of phosphates. The mass spectrum will also reveal the presence of any intermediate phosphorylated forms.

Electrospray ionization mass spectrometry (ESMS) and matrix assisted laser desorption ionization time of flight (MALDITOF) mass spectrometry may both be used in principle (42), although for molecules the size of linker histones the former gives more accurate values (typical mass accuracy 0.01%; ± 1 Da on 10 kDa). For the electrospray method, the best results are obtained if the sample is first purified by reversed-phase HPLC using volatile solvents (*Protocol 15*), dried down, and redissolved (e.g. in 50% acetonitrile–1% acetic acid or 50% methanol–1% acetic acid at ~ 5–25 pmol/μl) for injection (10 μl; flow 4 μl/min). Electrospray mass spectra of spermatid and sperm H1 are shown in *Figure 5*; the calculated masses show that the spermatid H1 contains a mixture of species with 7, 8, and 9 phosphates (the 8-phosphate form being the major component).

The MALDITOF technique requires only 1–10 pmol of sample and, unlike

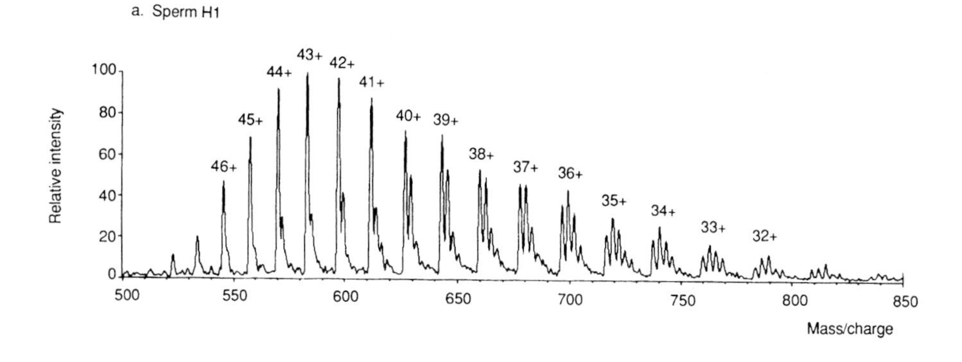

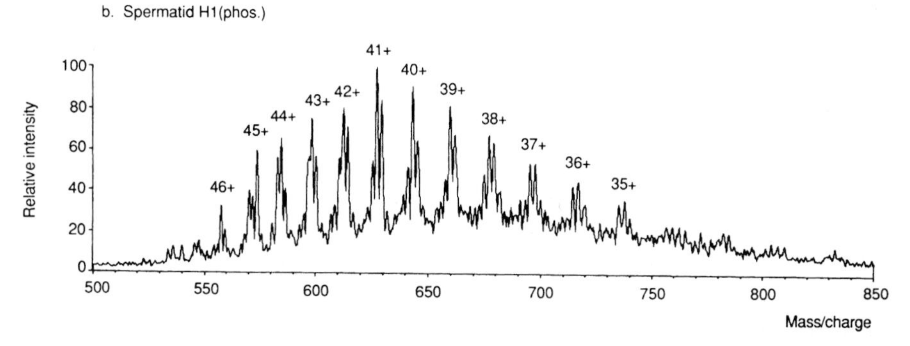

Figure 5. Demonstration of the number of phosphorylation sites in sea-urchin spermatid H1 by electrospray ionization mass spectrometry. Electrospray mass spectra for *P. miliaris* sperm and spermatid H1 recorded on a VG BioQ instrument. (a) Sperm H1. The spectrum indicates a single component of molecular weight 24 942.5 Da, with ~98 Da adducts (possibly sulfate) whose relative intensity is less for the more highly charged ions. (b) Spermatid H1. The peaks are triplets (seen most clearly for the 41+, 40+, and 39+ ions) arising from three components of molecular weight 25 505.1, 25 582.9, and 25 665.5 Da (all ± 2.5 Da), corresponding to H1 carrying 7, 8, and 9 phosphates, respectively (the 8-phosphate species being the most abundant). (Note that the deconvoluted spectrum (computer generated) of an electrospray mass spectrum of a single protein gives a single molecular mass from the population of multiply charged ions (42, 43), and this appears as a single peak in the data output. The deconvoluted spectrum of the spermatid H1 would thus have three peaks, the largest in area corresponding to the major component.) (From ref. 44, with permission.)

electrospray ionization, can tolerate some salts up to 0.1 M. The sample is dried onto a metal plate in the presence of a UV light absorbing matrix (e.g. α-cyano-4-hydroxycinnamic acid or sinapinic acid) and then salt and buffer components are washed away. Nonetheless, better results are often obtained if the sample is HPLC-purified as for ESMS, to disrupt possible sample-salt adducts which might confuse the mass analysis. Mass accuracy is typically 0.02–0.05%.

6.3 Determination of the positions of the phosphorylated residues in the protein sequence

Methods for the determination of phosphorylation sites in proteins form the topic of half a dozen chapters in a volume on protein phosphorylation (45) which provide a wealth of experimental detail and alternative procedures.

The amino-acid sequences of many H1 subtypes and variants is known (33). Identification of the sites of phosphorylation is perhaps most conveniently achieved by sequence analysis of ^{32}P-labelled peptides produced by the proteolytic digestion of *in vivo* ^{32}P-labelled histones. The radiolabelled peptides can be isolated by HPLC (cf. ref. 46), or eluted from a peptide map (loaded with about 1 nmol of digest) run on a silica-gel plate (8) using methods described elsewhere (47). A typical peptide map (of the N-terminal half of spermatid H1) and autoradiograph of phosphorylated peptides is shown in *Figure 6* (see also ref. 48).

Whether or not the phosphorylated histone is radiolabelled, phosphopeptides may be selected on iron-chelation columns (IDA Sepharose, Sigma) (see ref. 49 and references therein), or possibly identified from a slightly shorter retention time on a reversed-phase HPLC column compared with the unphosphorylated counterparts, due to increased hydrophilicity. In a systematic approach (50), aliquots of every HPLC-fractionated peak from a proteolytic digest of the phosphorylated protein were acid-hydrolysed; the liberated amino acids were modified at their amino groups by treatment with dabsyl chloride and the phosphoamino-acid derivatives identified by capillary electrophoresis (which will, however, not be available in all laboratories). In a similar vein, aliquots could be subjected to partial (3 h) acid hydrolysis and these examined for phosphoserine and phosphothreonine; if radioactive, the hydrolysates could be screened by thin-layer chromatography. More simply, provided the instrumentation is available, coupled HPLC/MS allows the HPLC peaks from a proteolytic digest to be fed directly into a mass spectrometer and their masses measured. If the sequence of the protein is known, the measured masses can be compared with the predicted peptide masses and components with predicted/observed mass differences of 80 Da (or multiples thereof) identified as phosphorylated. If they contain a single phosphorylatable residue the task is complete, otherwise sequence analysis is carried out on the relevant HPLC-fractionated component(s) as described below.

The positions of phosphorylated serine or threonine residues may be determined by automated amino-acid sequencing (e.g. using the Applied Biosystems Model 477A Pulsed-Liquid Sequencer with an on-line Model 120A PTH Amino Acid Analyser). Ideally, 200–500 pmol of peptides would be available, but 20–50 pmol should be sufficient. If the peptide is ^{32}P-labelled, the position of the phosphate can be determined by extracting the sequencer support after each cycle that gives Ser- or Thr-related products and counting for radioactivity (51), because when the phosphorylated residue reaches the N-terminal position the phosphate undergoes β-elimination. However, it may

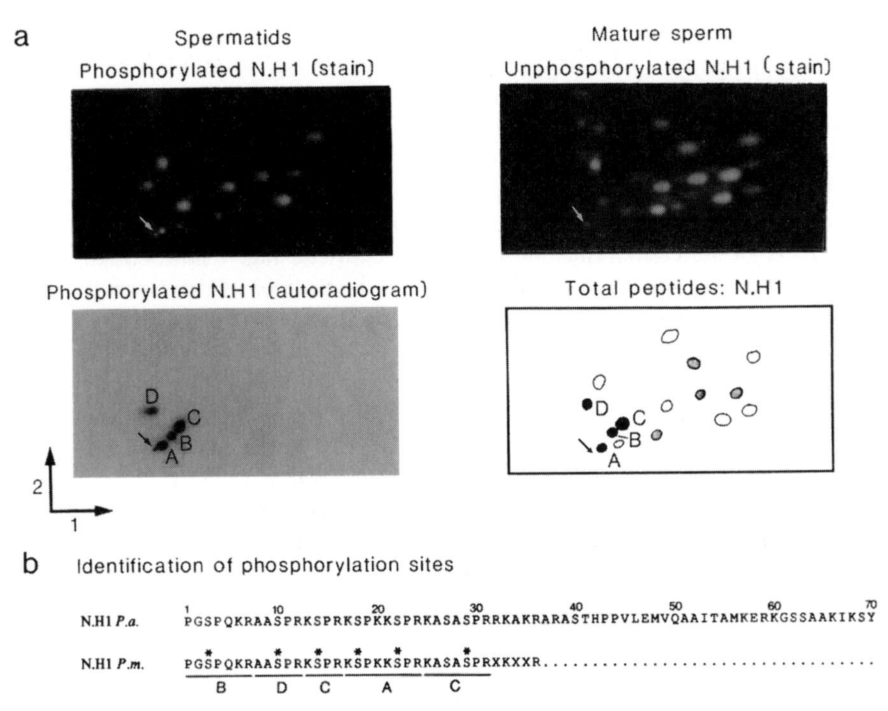

Figure 6. Identification of ^{32}P-labelled phosphorylated peptides by peptide mapping. Mapping of phosphorylation sites in the N-terminal domain of spermatid H1. (a) The N-terminal chymotryptic fragments (N.H1) from radiolabelled spermatid H1 and unlabelled sperm H1 were reductively methylated (at lysine side-chains, to limit subsequent tryptic cleavage to arginyl bonds), digested with trypsin, and the tryptic peptides were analysed by two-dimensional peptide mapping (47). Upper panels, fluorescamine-stained maps; lower left panel, autoradiogram (spermatid N.H1). First dimension (1), electrophoresis at pH 6.5 (10% pyridine, 0.5% acetic acid (v/v) in water); second dimension (2), chromatography (butan-1-ol/acetic acid/water/pyridine (15:3:12:10, v/v)). The origin is indicated by an arrow. The schematic diagram shows the positions of unphosphorylated peptides that become phosphorylated in spermatids (shaded), and their phosphorylated counterparts (coloured black). (The former stain strongly with fluorescamine in the map of unphosphorylated N.H1 but are weak in the map of phosphorylated N.H1; the latter appear only in the map of phosphorylated N.H1.) Peptides A–D were identified by sequence analysis or from their amino-acid composition and assigned within the N-terminal sequence of *P. miliaris* H1 (phosphorylation sites marked with asterisks) as shown in (b). (The sequence of the first 36 amino acids of *P. miliaris* H1 (*P.m.*) was determined on the intact unphosphorylated protein (150 pmol) and was identical, at least for the first 31 residues, with the sequence of *P. angulosus* H1 (*P.a.*) shown for comparison.) (From ref. 8, with permission.)

be difficult to recover the phosphate quantitatively from the support; this can lead to carry-over of radioactivity which might be a problem if there are additional serine or threonine residues in the sequence close to the first. Extraction is improved by using, for example, the more hydrophobic polyvinylidene difluoride (PVDF) as the support instead of polybrene-coated glass fibre, and by the introduction of, for example, methanol washes before the usual butyl chloride extraction of the anilinothiazolinone (ATZ) amino acid from the membrane. If the sample is covalently attached to a membrane (e.g. a Sequelon PVDF support; PerSeptive Biosystems), the phosphate may be extracted with neat trifluoroacetic acid.

Alternative methods for the identification of phosphorylated serine (or threonine) residues by automated sequencing can be applied whether or not a peptide is ^{32}P-labelled. They both make use of the fact that for serine, for example, the precursor (phenylthiocarbamyl derivative) of the phenylthiohydantoin (PTH) of the serine residue produced during Edman degradation undergoes partial side-chain dehydration by β-elimination to give the dehydroalanine derivative, and this largely forms an adduct with dithiothreitol (DTT) present in the solution. If the serine is phosphorylated, β-elimination is promoted and the ratio of the PTH derivative of the dehydroalanine–DTT adduct to PTH-serine increases (52). The method also works for phosphothreonine (53). The conditions of sequence analysis may be optimized to enhance β-elimination, e.g. by increasing the temperature of the coupling reaction in the presence of triethylamine (8). In a typically optimized case, the ratio of PTH-serine to PTH-dehydroalanine plus its DTT adduct was ~0.05:1 for a phosphorylated serine (compared with 0.4–0.5:1 for the unmodified residue).

The second method, widely used, that does not require the peptide to be radioactive for identification of phosphoserine, also starts with β-elimination. The product is then converted, by the addition of ethanethiol, to *S*-ethylcysteine whose PTH derivative is readily identified by sequence analysis (54). The method also works for phosphothreonine, provided the conditions are modified (50). Care is needed with –SPKK– motifs because phosphoseryl residues in the sequence –SP– are resistant to β-elimination and need more drastic reaction conditions (55).

Methods for peptide sequencing by tandem mass spectrometry (MS/MS) continue to be developed (43). (The first mass analyser picks out one peptide species of interest from a complex mixture and the second fragments it, mainly in the amides of the peptide backbone.) This approach can identify phosphorylated peptides and sites of phosphorylation directly, starting with a complex mixture, and will doubtless be widely exploited for this purpose, where instrumentation and expertise allow. In less specialized laboratories, the methods described above are perfectly adequate, especially for histones, which unlike the many biologically important phosphoproteins involved in the various cell-signalling pathways, for example, are not in particularly short supply.

Acknowledgements

I thank Dr Len Packman for helpful comments on section 6. Work in this laboratory is supported by the Wellcome Trust and the Biotechnology and Biological Sciences Research Council.

References

1. Thomas, J. O. and Butler, P. J. G. (1977). *J. Mol. Biol.* **116**, 769.
2. Wassarman, P. M. and Kornberg, R. D. (ed.) (1989). *Methods in enzymology*. Vol. 170. Academic Press, New York.
3. Hewish, D. R. and Burgoyne, L. A. (1973). *Biochem. Biophys. Res. Commun.* **52**, 504.
4. Weintraub, H., Palter, K., and Van Lente, F. (1975). *Cell* **6**, 85.
5. Thomas, J. O., Rees, C., and Butler, P. J. G. (1986). *Eur. J. Biochem.* **154**, 343.
6. Spadafora, C., Bellard, M., Compton, J. L., and Chambon, P. (1976). *FEBS Lett.* **699**, 281.
7. Green, G. R. and Poccia, D. L. (1988). *Biochemistry* **27**, 619.
8. Hill, C. S., Packman, L. C., and Thomas, J. O. (1990). *EMBO J.* **9**, 805.
9. Thompson, R. J. (1973). *J. Neurochem.* **21**, 19.
10. Pearson, E. C., Butler, P. J. G., and Thomas, J. O. (1983). *EMBO J.* **2**, 1367.
11. Thomas, J. O. and Thompson, R. J. (1977). *Cell* **10**, 633.
12. Pearson, E. C., Bates, D. L., Prospero, T. D., and Thomas, J. O. (1984). *Eur. J. Biochem.* **144**, 353.
13. Noll, M., Thomas, J. O., and Kornberg, R. D. (1975). *Science* **187**, 1203.
14. Caron, F. and Thomas, J. O. (1981). *J. Mol. Biol.* **146**, 513.
15. Thomas, J. O. and Rees, C. (1983). *Eur. J. Biochem.* **134**, 109.
16. Lutter, L. C. (1978). *J. Mol. Biol.* **124**, 391.
17. Kornberg, R. D., LaPointe, J. W., and Lorch, Y. (1989). In *Methods in enzymology* (ed. P. M. Wassarman and R. D. Kornberg), Vol. 170, p. 3. Academic Press, New York.
18. Bates, D. L., Butler, P. J. G., Pearson, E. C., and Thomas, J. O. (1981). *Eur. J. Biochem.* **119**, 469.
19. Butler, P. J. G. and Thomas, J. O. (1980). *J. Mol. Biol.* **140**, 505.
20. Noll, H. and Noll, M. (1989). In *Methods in enzymology* (ed. P. M. Wassarman and R. D. Kornberg,), Vol. 170, p. 55. Academic Press, New York.
21. Panyim, S., Jensen, R. H., and Chalkley, R. (1968). *Biochim. Biophys. Acta* **160**, 255.
22. Harborne, N. and Allan, J. (1983) *FEBS Lett.* **155**, 89.
23. Bates, D. L. and Thomas, J. O. (1981). *Nucl. Acids Res.* **9**, 5883.
24. Thomas, J. O. and Kornberg, R. D. (1975). *Proc. Natl Acad. Sci. USA* **72**, 2626.
25. Thomas, J. O. and Kornberg, R. D. (1978). *Methods Cell Biol.* **18**, 429.
26. Laemmli, U. K. (1970). *Nature* **227**, 680.
27. Allan, J., Staynov, D. Z., and Gould, H. (1980). *Proc. Natl Acad. Sci. USA* **77**, 885.
28. Brand, S. H., Kumar, N. M., and Walker, I. O. (1981). *FEBS Lett.* **133**, 63.
29. De Petrocellis, L., Quagliarotti, G., Tomei, L., and Geraci, G. (1986). *Eur. J. Biochem.* **156**, 143.
30. Clark, D. J. and Thomas, J. O. (1986). *J. Mol. Biol.* **187**, 569.
31. Clark, D. J., Hill, C. S., Martin, S. R., and Thomas, J. O. (1988). *EMBO J.* **7**, 69.
32. Simon, R. H. and Felsenfeld, G. (1979). *Nucl. Acids Res.* **6**, 689.
33. Wells, D. and Brown, D. (1991). *Nucl. Acids Res.* **19**, (Suppl.) 2173.
34. Camerini-Otero, R. D., Sollner-Webb, B., and Felsenfeld, G. (1976). *Cell* **8**, 333.
35. Ajiro, K., Borun, T. W., and Cohen, L. H. (1981). *Biochemistry* **20**, 1445.
36. Lennox, R. W. and Cohen, L. H. (1988). In *Chromosomes and chromatin* (ed. K. W. Adolph), Vol. 1, p. 33. CRC Press, Boca Raton, FL.

37. Paulson, J. R. (1980). *Eur. J. Biochem.* **111**, 189.
38. Panyim, S. and Chalkley, R. (1969). *Biochem. Biophys. Res. Comm.* **37**, 1042.
39. Zweidler, A. (1978). *Methods Cell Biol.* **17**, 223.
40. Lennox, R. W. and Cohen, L. H. (1989). In *Methods in enzymology* (ed. P. M. Wassarman and R. D. Kornberg), Vol. 170, p. 532. Academic Press, New York.
41. Paulson, J. R., Mesner, P. W., Delrow, J. J., Mahmoud, N. N., and Ciesielski, W. A. (1992). *Anal. Biochem.* **203**, 227.
42. Geisow, M. (1992). *Trends Biotechnol.* **10**, 432.
43. Mann, M. and Wilm, M. (1995). *Trends Biochem. Sci.* **20**, 219.
44. Hill, C. S., Rimmer, J. M., Green, B. N., Finch, J. T., and Thomas, J. O. (1991). *EMBO J.* **10**, 1939.
45. Hunter, T. and Sefton, B. M. (ed.) (1991). *Methods in enzymology*, Vol. 201. Academic Press, San Diego.
46. Thomas, J. O. and Wilson, C. M. (1986). *EMBO J.* **5**, 3531.
47. Thomas, J. O. (1989). In *Methods in enzymology* (ed. P. M. Wassarman and R. D. Kornberg), Vol. 170, p. 369. Academic Press, New York.
48. Ajiro, K., Borun, T. W., Shulman, S. D., McFadden, G. M., and Cohen, L. H. (1981). *Biochemistry* **20**, 1454.
49. Miller, A. D., Packman, L. C., Hart, G. J., Alefounder, P. R., Abell, C., and Battersby, A. R. (1989). *Biochem. J.* **262**, 119.
50. Meyer, H., Eiserman, B., Heber, M., Hoffmann-Posorske, E., Korte, H., Weigt, C., Wegner, A., Hutton, T., Donella-Deana, A., and Perich, J. W. (1993). *FASEB J.* **7**, 776.
51. Roach, P. J. and Wang, Y. (1991). In *Methods in enzymology* (ed. T. Hunter and B. M. Sefton), Vol. 201, p. 200. Academic Press, San Diego.
52. Meyer, H. E., Hoffmann-Posorske, E., Korte, H., and Meyer, L. M. G. (1986). *FEBS Lett.* **204**, 61.
53. Dedner, N., Meyer, H. E., Ashton, C., and Wildner, G. F. (1988). *FEBS Lett.* **236**, 77.
54. Meyer, H. E., Hoffmann-Posorske, E., and Heilmeyer, J., L.M.G. (1991). In *Methods in enzymology* (ed. T. Hunter and B. M. Sefton), Vol. 201, p. 169. Academic Press, San Diego.
55. Cohen, P., Gibson, B. W., and Holmes, C. F. B. (1991). In *Methods in enzymology* (ed. T. Hunter and B. M. Sefton), Vol. 201, p. 153. Academic Press, San Diego.
56. Loening, U. E. (1967). *Biochem. J.* **102**, 251.

2

Extraction and separation of core histones and non-histone chromosomal proteins

ALAN W. THORNE, PETER D. CARY, and
COLYN CRANE-ROBINSON

1. Introduction

The participation of the nucleosome in DNA processing events such as transcriptional activation is now well established (see Editor's Introduction). It has been shown, for example, that certain transcription factors can access their recognition sites within a nucleosome alone or with the help of the SWI/SNF complex. The involvement of core histone modifications, particularly the acetylation of lysine side chains in the N-terminal tails of histones H3 and H4, has been studied both by yeast genetics and *in vitro* methods. The *in vitro* recapitulation of key events in the reprogramming of nucleosomes consequent upon enzymatic modifications and the binding of transcription factors, RNA polymerase, chromatin assembly factors, or other proteins involved in DNA processing, requires the effective reconstitution of nucleosomes, often on specific DNA sequences and with a defined composition of post-translational histone modifications. The ideal prerequisite is to have precisely the correct histones to hand for reconstitution. For some experiments, simple transfer of histone octamers from mono- or oligonucleosomal donor templates has been sufficient and in the case of hyperacetylated nucleosomes those from butyrate-treated HeLa nuclei have been especially useful. In the future, however, a more precisely defined modification pattern of the core histones will be required (be it acetylation, ubiquitination, methylation, or phosphorylation) and then one must resort to initial separations of the component histones followed by nucleosomal reconstitution. The same is true if *in vitro* experiments to mimic *in vivo* mutations are envisaged and mutant histones have to be heterogeneously expressed and incorporated into nucleosomes. For these reasons, the separation of individual core histones, core histone pairs, and even octamers is not an art that has had its day. As regards linker histones, the subject is somewhat simpler in that monomer protein is always the required component of reconstitution, although in this case concerns remain regarding irreversible denaturation of structural elements in the N- and C-terminal domains following acid-extraction methods, and the lack of effective assays for linker-histone reconstitution make this a rather difficult problem. Nevertheless, the inclusion of linker histones, post-translationally modified or not, into reconstituted chromatin is an essential step for a complete understanding of chromatin repression and expression.

Section 3 describes methods of histone extraction from nuclei of calf thymus, chicken erythrocyte and mammalian cells in culture. Since core histones differ little between these three sources, if unmodified histones are required, erythrocytes from mature hens represent the most convenient and readily available source (see Chapter 1). If mammalian core histones are specifically required, then calf thymus (a tissue with a large nuclear to cytoplasmic ratio) is the most convenient source. If modified histones are required, an active tissue, such as rat liver or cells in culture, is the most convenient, and butyrate treatment can be used to provide acetylated histones. Proteolytic degradation is easier to control for cells in culture and this is therefore the procedure described. If linker histones are required, then tissue-specific variation will determine the source. The convenience of perchloric acid extraction for linker-histone extraction is considerable and a special section is devoted to this technique. Despite the fact that perchloric acid denatures linker histones, it does not modify them and they can be completely renatured. Trichloroacetic acid precipitation, on the other hand, should never be used for any histones since side-chain acylation can occur and the reversibility of the denaturation is poor. For core histones, renaturation of acid-extracted proteins is less efficient, particularly from the four individual core histones. Thus if H3/H4 tetramers, H2A/H2B dimers, or H3/H4/H2A/H2B octamers are to be prepared from acid-extracted histones (typically by dialysis from high-salt conditions), a purification step of the final species will be required. If nucleosomes or core particles are required, their formation to some extent represents a purification, but a further polishing step will nevertheless be necessary. Alternatively, native histone octamers can be prepared by capture of chromatin fragments on hydroxyapatite and elution with a high-salt buffer. This is a particularly convenient method if an unaltered core histone octamer is required, and the method has the additional merit that undenatured linker histones and high-mobility group (HMG) proteins can be extracted by prior elution with lower salt concentrations.

The fractionation methods to be employed depend very much on the scale of the extraction. The soft-gel methods described for the individual histones allow large quantities to be prepared (up to 1 g) but they are somewhat slow and laborious. Reverse-phase methods are much quicker and more convenient but generate less product (up to 1 mg). These considerations apply whether unmodified histones are to be separated or the aim is to fractionate modified subspecies of the individual histones.

Three types of gel electrophoresis are described. Laemmli SDS gels are used if post-translational modification is not an issue. If it is, then acetic acid/urea gels are used, with or without Triton X-100. The addition of this detergent serves to markedly retard histone H2A, so that the pattern of H2B species is more clearly revealed. Primary sequence variants, in particular of histone H3, are also separated in the presence of Triton X-100, which for certain sources, e.g. human, complicates the pattern considerably, so a gel without detergent should be used initially.

2. Preparation of nuclei from different tissues

Whilst histones can be acid-extracted by homogenization of whole tissue in various acids or salt solutions, cleaner preparations of histones are obtained if the nuclei of the source tissue are isolated first.

2.1 Preparation of nuclei from calf thymus

The preparation method given below is adapted from Panyim *et al.* (1), by lowering the $MgCl_2$ concentration to 6 mM and the inclusion of sodium butyrate and protease inhibitors. Since homogenization efficiency varies depending on the blender used, times and speed of homogenization are determined by examination of the nuclei by phase-contrast microscopy throughout the preparation.

Protocol 1. Preparation of nuclei from calf thymus

Equipment and reagents

- Grinding buffer: 0.25 M sucrose, 10 mM Tris–HCl, pH 7.4, 6 mM $MgCl_2$, 10 mM Na butyrate, 0.1 mM phenyl methyl sulfonyl fluoride (PMSF), 0.1 mM benzamidine
- Wash buffer A: as above including 0.2% w/v Triton X-100
- Domestic nylon sieve
- Sucrose cushion: 30% sucrose containing 10 mM Tris–HCl, pH 7.4, 6 mM $MgCl_2$, 10 mM Na-butyrate, 0.1 mM PMSF, 0.1 mM benzamidine
- Waring blender
- Refrigerated centrifuges fitted with a swing-out rotor (Sorvall RC-3B, RC-5B) Rotors H6000A, HB4

Method

1. Collect calf thymus glands immediately after slaughter and freeze in liquid nitrogen. Frozen thymus can be stored for several months at −80 °C without deterioration of the histones.
2. Remove the bulk of any connective tissue, blood vessels, and fat deposits by dissection from partially thawed thymus. Cut the thymus into small (< 2 cm) pieces.
3. Homogenize 40–50 g of thymus in 700 ml of grinding buffer in a blender at low speed for 1 min and high speed for 2 min, or until a homogenous suspension is just achieved.

Protocol 1. *Continued*

4. Filter the homogenate through a domestic nylon sieve to remove any connective tissue.
5. Centrifuge at 2000 *g* for 10 min at 4°C using a swing-out rotor. Pour off the supernatant. Resuspend the pellet in wash buffer A (containing Triton X-100) and homogenize for 3 min at low speed. Centrifuge as above.
6. Repeat the homogenization and centrifugation steps until the released nuclei are free of cytoplasmic tags, as judged by phase-contrast microscopy, and the supernatant is clear (usually three cycles).
7. Resuspend the final nuclear pellet in grinding buffer (no Triton X-100) and centrifuge as above. **NB** Nuclei may be stored at this stage at −80°C by resuspending in a minimum volume of grinding buffer (no Triton X-100) and mixing with an equal volume of 80% glycerol in the same buffer.
8. Final purification is achieved by pelleting the nuclei through a sucrose cushion. Resuspend the pellet in grinding buffer to approximately 4 mg DNA/ml and pellet the nuclei through a 30% sucrose cushion in a swing-out rotor. Overlay 2 ml nuclear suspension over a 5 ml cushion and centrifuge at 2400 *g* for 5 min in a Sorvall HB4 swing-out rotor at 4°C. Discard the supernatant. If stored (frozen) nuclei are to be used the glycerol must be removed prior to overlaying on the sucrose as follows: thaw the nuclei and dilute by adding at least three volumes of grinding buffer, centrifuge at 2000 *g* for 10 min at 4°C, resuspend to approximately 4 mg DNA/ml and proceed as above.

2.2 Preparation of nuclei from mammalian cells

Protocol 2. Preparation of nuclei from mammalian culture cells

The following method is used to prepare nuclei from a number of human cell lines such as K562, HL60, and HeLa in spinner flask culture (but see Chapter 5, *Protocol 1*).

Equipment and reagents

- PBS-butyrate: 135 mM NaCl, 2.5 mM KCl, 8 mM Na_2HPO_4, 1.5 mM KH_2PO_4, 10 mM Na-butyrate
- Cell lysis buffer: 250 mM sucrose, 10 mM Tris–HCl, pH 7.4, 10 mM Na-butyrate, 4 mM $MgCl_2$, 0.1 mM PMSF, 0.1 mM benzamidine, 0.1% (w/v) Triton X-100
- Sucrose cushion: 30% (w/v) sucrose in wash buffer C
- Wash buffer C: 250 mM sucrose, 10 mM Tris–HCl, pH 7.4, 10 mM Na-butyrate, 4 mM $MgCl_2$, 0.1 mM PMSF, 0.1 mM benzamidine
- Storage buffer: 80% glycerol in wash buffer C
- Dounce homogenizer with B pestle
- Refrigerated centrifuges with swing-out rotor (as *Protocol 1*).

Method

1. Harvest the cells from spinner flask culture at densities between 5 and 10×10^5 cells/ml by centrifugation at 2500 *g* for 20 min at 4 °C.
2. Resuspend the cell pellet in ice-cold PBS-butyrate and repeat the centrifugation step as above.
3. Resuspend the pellet in cell lysis buffer containing 0.1% (w/v) Triton X-100 and lyse with 10 strokes of the Dounce homogenizer.
4. Pellet the nuclei by centrifugation at 2000 *g* for 10 min at 4 °C. Discard the supernatant and resuspend the pellet in about 5 ml of wash buffer C.
5. Overlay 2 ml of the nuclear suspension onto 5 ml of sucrose cushion and centrifuge at 2400 *g* for 5 min at 4 °C in a swing-out rotor as described for calf thymus nuclei.
6. Remove all the supernatant and resuspend the pellet in a minimum volume of wash buffer C. Nuclei may be stored at this stage by adding an equal volume of storage buffer and freezing at −80 °C.

3. Histone extraction methods

3.1 Acid extraction of histones from isolated nuclei (all sources)

Histones can be acid-extracted directly from a nuclear suspension provided there is no sucrose or glycerol present in the nuclear wash buffer. These are removed by resuspending and centrifuging the nuclear pellet in the appropriate wash buffer, without sucrose or glycerol, immediately prior to extraction.

Protocol 3. Acid extraction of histones from isolated nuclei (all sources)

Equipment and reagents

- Appropriate nuclear wash buffer without sucrose
- Refrigerated centrifuge (Sorvall RC-5B) with fixed angle rotor (SS34)

Method

1. Resuspend the nuclear pellet in a minimum volume of the appropriate nuclear wash buffer, omitting the sucrose, and stir on ice.
2. Add an equal volume of ice-cold 0.8 M HCl. Stir on ice for 2 hours.
3. Remove acid-insoluble material by centrifugation at 12 000 *g* for 10 min at 4 °C. Retain the histone-containing supernatant and the pellet for re-extraction.

Protocol 3. *Continued*

4. Disperse the pellet, by homogenization if necessary, in a further volume of ice-cold 0.4 M HCl and stir for a further 2 h on ice.
5. Centrifuge the suspension as above and combine the supernatant with that retained earlier.
6. Precipitate the histones by adding eight volumes of acetone and leave at −20°C overnight.
7. Recover the histones by centrifugation at 2000 *g* for 10 min at 4°C.
8. Remove excess salts at this stage by washing the histone pellets twice with freshly prepared acetone/100 mM HCl (10:1) at room temperature. Use at least 10 pellet volumes of acetone/HCl for each wash. Recover the histone precipitate by centrifugation at 2000 *g* for 10 min at room temperature.
9. Wash the histone pellets three times with at least 10 pellet volumes of dry acetone at room temperature, centrifuging as above.
10. Dry the final pellet under vacuum at room temperature and store at −20°C.

Protocol 4. Acid extraction from lysed nuclei

This method gives a more efficient extraction of histones but is more laborious.

Equipment and reagents

- Appropriate nuclear wash buffer
- 0.25 mM EDTA
- Magnetic stirrer(Bibby H501)
- Refrigerated centrifuge (Sorvall RC-5B) with fixed angle rotor (SS34)

Method

1. Resuspend the nuclei in a *minimum* volume of the appropriate nuclear wash buffer.
2. Add the nuclear suspension dropwise to at least 20 volumes of rapidly stirring ice-cold 0.25 mM EDTA and leave stirring for 20 min to lyse the nuclei. Centrifuge the lysed nuclei for 30 min at 14 000 *g* to recover the chromatin pellet.
3. Disperse the chromatin pellet (a gel) in ice-cold 0.25 mM EDTA using a *minimum* of homogenization to complete the lysis, then repeat the centrifugation as above.
4. Homogenize the chromatin gel in 10 volumes of ice-cold H_2O to fully disperse the pellet and then add concentrated HCl to a final concentration of 0.4 M. Stir on ice for 2 h.
5. Follow steps 3–10 of *Protocol 3*.

3.2 Extraction of histones from chromatin using salt

Salt extraction procedures are advantageous since the histones can be prepared in pure form with little contaminating non-histone proteins and the core histone octamer can be readily isolated. The method relies on the differential affinities of histones and non-histone proteins for DNA in varying salt conditions. Non-histone chromosomal proteins (HMGs, etc.) can be removed by washing chromatin or nuclei in 0.35 M NaCl-containing buffers. H1 (and H5) are released by increasing the salt levels to above 0.6 M NaCl, and finally the core histones are extracted as octamers in buffers containing 2 M NaCl. Whilst the different fractions can be extracted by direct treatment of chromatin or nuclei it is more convenient to immobilize the DNA by binding to hydroxyapatite and washing through with the various salt solutions. The following method (2) elutes both non-histone proteins and H1/H5 in a single step followed by the core histone octamer.

Protocol 5. Salt extraction of histones from chromatin

Equipment and reagents

- Bio-Rad Econopac 120 × 15 mm column
- Bio-Rad hydroxyapatite (Bio-Gel HTP)
- 0.65 M NaCl–phosphate buffer: 50 mM phosphate, pH 6.8, 0.65 M NaCl, 0.1 mM PMSF, 0.1 mM benzamidine (add protease inhibitors immediately prior to use)
- Appropriate nuclear wash buffer
- 2.5 M NaCl–phosphate buffer: 50 mM phosphate, pH 6.8, 2.5 M NaCl, 0.1 mM PMSF, 0.1 mM benzamidine (add protease inhibitors immediately prior to use)
- Roller (SRT1 Stuart Scientific)
- Magnetic stirrer (Bibby H501)

Method

1. Prepare nuclei (total 6 mg DNA) as described earlier and resuspend in a minimum volume of the appropriate nuclear wash buffer.
2. Add dropwise to 10 ml ice-cold 0.65 M NaCl–phosphate buffer under gentle stirring. The nuclei should start to lyse immediately and the solution become slightly viscous. Roll gently at 4 °C for 20 min.
3. Add 4 g dry hydroxyapatite to form a paste and transfer into the chromatography column on top of 0.25 g hydroxyapatite equilibrated in the above buffer. The total resin volume should be about 10 ml.
4. Allow the resin to settle and elute the H1 with 50 ml of 0.65 M NaCl–phosphate buffer. H1 and other proteins may be recovered from this eluant by dialysing against H_2O prior to freeze drying.
5. Elute the core histones (as octamers) by passing 50 ml of 2.5 M NaCl–phosphate buffer through the column. The octamers can be concentrated by ultrafiltration using an Amicon concentrator with a YM10 membrane and stored at −20 °C.

The method can be modified to remove HMGs and other non-histone proteins prior to extraction of H1/H5 by the inclusion of a 0.35 M NaCl elution step. In this case, nuclei are lysed by repeated resuspension and centrifugation in a low ionic strength buffer (e.g. 0.25 mM EDTA, 10 mM Tris–HCl, pH 7.5, 10 mM Na-butyrate, 0.1 mM PMSF, 0.1 mM benzamidine). If necessary, chromatin pellets can be dispersed by sonication prior to mixing with the hydroxyapatite.

4. Gel electrophoresis of histones

SDS-PAGE analysis is performed in 15% acrylamide gels after the method of Laemmli (3). Histones and other proteins can be precipitated from P60 fractions using ammonium reineckate (see *Protocols 6–8*). However, since the salt and pH of the P60 buffer is not a problem for such SDS gels, column fractions can be loaded directly. Histone markers are then loaded in the same volume of column buffer.

Acetic acid–urea PAGE is performed according to the method of Panyim and Chalkley (4) and is used to monitor the presence and level of post-translational charge modifications to the histones, e.g. acetylation. We use gels with a final composition of: 15% acrylamide, 0.41% bisacrylamide, 6.25 M urea, 0.9 M acetic acid. These gels are very sensitive to the presence of salt in the sample, so reineckate precipitation and acetone/100 mM HCl (10:1) washing should be performed on all samples to ensure salt removal. Samples are dissolved in fresh 8 M urea, 10 mM mercaptoethanol for at least 30 min prior to the addition of acetic acid to 5% for loading. These gels have the disadvantage that the acetylated subfractions of histones H2A and H2B overlap and the Triton X-100 system must be used to resolve these.

Acetic acid–urea–Triton X-100 PAGE is performed after the method of Bonner *et al.* (5). Triton X-100 (a non-ionic detergent) binds to and retards the migration of the histones, in particular the more hydrophobic H2A. The pattern in the H2B region of the gel is very much simplified. However, the primary sequence variants of H2A and H3 also separate in this gel system. Electrophoresis in a second dimension (e.g. using SDS-PAGE) is therefore often required for unambiguous assignment of all bands.

Protocol 6. Histone precipitation with reineckate

Ammonium reineckate is very effective at precipitating histones and other proteins from dilute solution (e.g. 5 μg/ml) in buffers containing NaCl or urea.

Reagents

- Saturated solution of ammonium reineckate in 0.5 M HCl
- Refrigerated microfuge (Jouan mR22i)
- Vacuum dryer (Savant)
- Acetone (BDH Analar)

Method

1. Add an equal volume of the saturated reineckate solution to the solution of histone.
2. Leave on ice for 10 min.
3. Centrifuge at 12 000 *g* for 5 min at 4°C. Remove all the supernatant.
4. Resuspend the pellet by vortexing with 1 ml acetone/100 mM HCl (10:1) at room temperature.
5. Centrifuge at 12 000 *g* for 3 min at room temperature. Remove the supernatant by aspiration with a drawn-out Pasteur pipette taking care to avoid the pellet.
6. Repeat the wash and centrifugation steps 5 and 6.
7. Wash and centrifuge the pellet as above three times using dry acetone, and vacuum dry.

5. Histone fractionation by chromatographic methods

5.1 Large-scale separation of histone fractions

Large quantities of total or core histone extracts (20–1000 mg) can be separated into their principal subfractions by gel filtration chromatography using a combination of Bio-Gel P60 (modified from van der Westhuyzen *et al.* (6)) and P10 column chromatography. The buffer composition (principally the salt concentration) for the P60 column needs to be varied for optimal separation of histones from different organisms. The buffer conditions used below are suitable for histones from calf thymus. We have found that the optimal column lengths are about 1.2–1.4 m. Column diameter is selected to suit the quantities of histone to be separated. Histone is loaded up to a maximum of 15–20 mg/cm^2 cross-sectional area at a concentration of up to 50 mg/ml.

Protocol 7. P60 and P10 column preparation

Method

1. Swell the resin in the appropriate column buffer overnight at room temperature.
2. Thoroughly degas and remove fines prior to packing.
3. Pack large columns in the normal way using a packing funnel. Pack small columns (3–8 mm diameter) by drawing a suspended slurry into an inverted column under vacuum until just full, inverting the column, and then topping up in the usual way as the resin settles.

Protocol 7. *Continued*

4. After initial packing, pump buffer through the column overnight at a flow rate of up to 2.5 ml/h/cm^2 cross-sectional area of the column to ensure complete settling. If required, top the column up with resin and allow it to settle under flow prior to loading.

Protocol 8. Fractionation of total histone on Bio-Gel P60

Reagents

- Loading buffer 1: 8 M urea, 0.1 M DTT
- Elution buffer: 30 mM NaCl, 0.09% w/v formic acid, 0.02% Na-azide pH 2.5

Method

1. Dissolve total histone extracts at 50 mg/ml in freshly made loading buffer, overnight with gentle agitation (e.g. rolling) at room temperature to ensure complete reduction of disulfide bonds in histone H3 and complete dispersion of aggregates.
2. Load the column with the sample and elute at 2.5 ml/h/cm^2 using the elution buffer. Monitor the eluant at 230 nm.
3. Collect fraction sizes of 1/100th of the column volume.
4. Pool appropriate fractions and dialyse against 0.1% formic acid (3 changes 1:100, 48 h total) to remove salt and azide prior to lyophilization, or alternatively neutralize with NaOH, freeze-dry and then desalt either by dialysis or by using an appropriately sized Sephadex G-25 (or similar) column equilibrated in 0.1% formic acid. Individual histones or histone mixtures are stored in a lyophilized state at −20 °C.

Figure 1 shows a typical profile for calf thymus histones, H1 is eluted close to the void volume and is followed by H3, H2A, H2B a second H3 subfraction and finally H4. We have studied the effect of salt concentration in the elution buffer from 0 to 150 mM NaCl for calf thymus histones. As the salt level is increased the order of elution changes so that at 150 mM NaCl the order is H1, H2A, H3 with H2B and H4 co-eluting. Optimal separation of histones is achieved for calf thymus histones at 30 mM, conservative cuts from the peaks producing >95% purity of the pricipal histones. Separation of coeluting species, e.g. fractions containing H3 and H2A, can be achieved by chromatography on Bio-Gel P10 (see below). We routinely analyse aliquots of different fractions by 15% SDS-PAGE to ensure that fractions containing overlapping species are avoided. This is also used to identify fractions containing ubiquitinated histones uH2A and uH2B, which are also present on the leading half of the peaks containing H2A and H2B.

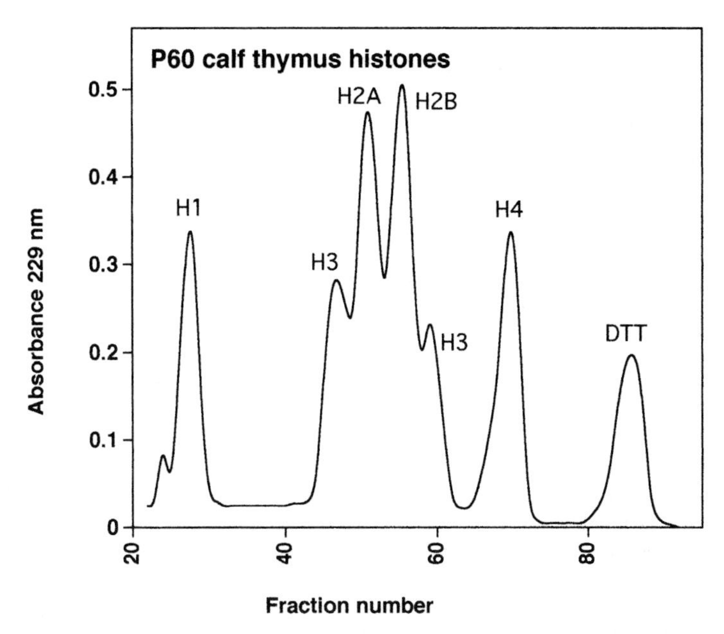

Figure 1. Bio-Rad P60 gel filtration of all 5 histones from calf thymus in 30 mM NaCl, 1200 × 25 mm, 0.02% Na-azide. Column dimensions: 0.09% w/v formic acid. Void volume at fraction 23. Flow rate: 2.5 cm/h. Total loading: 70 mg of acetone-precipitated histone in 2 ml.

Protocol 9. Bio-Gel P10 fractionation of calf thymus or HeLa H2A/H3

Equipment and reagents

- Loading buffer 1 (*Protocol 8*)
- 1.2 m P10 column (*Protocol 7*)
- 0.5% formic acid

Method

1. Prepare a 1.2 m P10 column of an appropriate diameter as described above (*Protocol 7*), equilibrated in 10 mM HCl, pH 2.0.
2. Dissolve the lyophilized (desalted) mixture of calf thymus H2A and H3 (or chicken erythrocyte H2B and H3) in 8 M urea, 0.1 M DDT loading buffer 1, *Protocol 8*) as for total histone.
3. Add concentrated HCl to 10 mM.
4. Load the column and elute with the above buffer at a flow rate of 2.5 ml/h/cm^2.
5. Monitor the eluant at 230 nm.
6. Collect fractions of size 1/100th column volume.

Protocol 9. *Continued*

7. Pool appropriate fractions and dialyse against 0.5% formic acid prior to lysophilization.
8. Store lyophilized samples at − 20 °C.

Figure 2 shows the separation of an H2A/H3 mixture derived from a P60 separation of HeLa histones.

5.2 Small-scale purification of histone fractions

Small quantities of histones (10–1000 μg) are most conveniently separated by chromatography on reverse-phase HPLC columns. A variety of carbon chain lengths (C-8, C-4, and C-3 and the propyl CN chemistry (Zorbax)) can be used to separate the main histone fractions. Buffers and elution conditions need to be optimized for each column type. We routinely perform a test separation by eluting total histone extracts from the reverse-phase column using a gradient from H_2O/ trifluoroacetic acid (TFA) to 70% acetonitrile/TFA using a rate of increase of 1%/min. TFA can be used at either 0.05% or 0.1% v/v. Retention times of the histone fractions are increased with the higher TFA

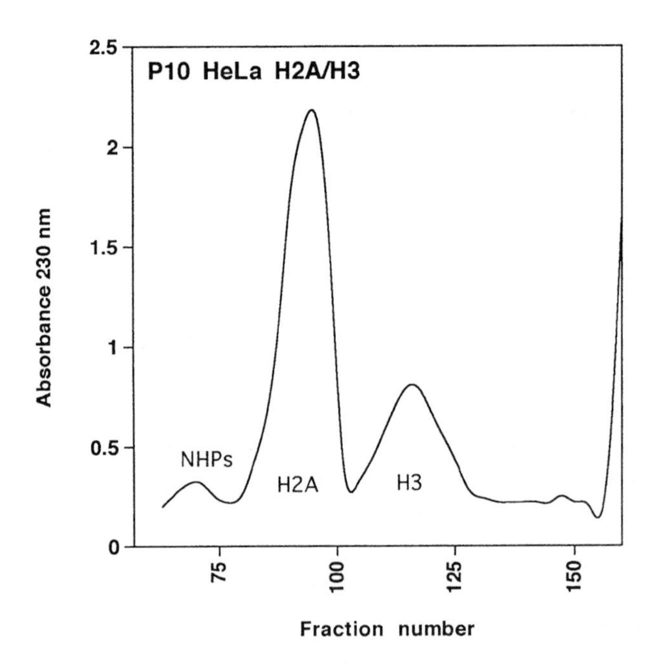

Figure 2. Bio-Rad P10 gel filtration of a HeLa cell histone H2A/H3 mixture resulting from a P60 fractionation run in 10 mM HCl. Column dimensions: 1470 × 35 mm.Void volume at fraction 67. Flow rate: 2.5 cm/h.Total loading: 120 mg of freeze-dried histone.

concentrations. The gradient shape is then modified to optimize resolution and minimize the time for separation. Typically this involves creating a shallow gradient for the H2A, H2B, and H4 region, since these histones elute close together, but a steeper gradient for the H3 region, since H3 elutes as a broad peak. TFA-containing buffers are convenient since the histones can be recovered by direct lyophilization. Using a CN column the principal histones, including H1 and H5, are separated in a single run, whilst on other columns, e.g. a C-8 column, H2B and H4 co-elute. This pair can be separated by rechromatography on the same C-8 column using a water/acetonitrile gradient containing acetic acid, $CaCl_2$, and methanol, rather than TFA. Histones are recovered by evaporation of the acetonitrile in a vacuum centrifuge followed by dialysis or precipitation using ammonium reineckate.

Protocol 10. Reverse-phase separation of core histones

A. *TFA buffer system*

Equipment and reagents

- Buffer A: 0.1% (v/v) trifluoroacetic acid (TFA) (Pierce) in H_2O
- Buffer B: 0.1% (v/v) TFA in acetonitrile (Rathburn)
- Zorbax SB300-CN (250 × 4.6 mm) reverse-phase column

Method

1. Use the following elution conditions for the Zorbax SB300-CN (250 × 4.6 mm) reverse-phase column: initial conditions: flow rate 1.4 ml/min (maintained throughout), 25% B, linear gradient 25–32% B over 10 min, linear gradient 32–37% B over 5 min, linear gradient 37–42% B over 20 min, linear gradient 42–55% B over 20 min. Retention times of principal chicken erythrocyte histones under these conditions are: H5, 13.5 min; H1, 15 min; H2B, 25 min; H2A, 27 min; H4 29.5 min; and H3, 44 min. When calf thymus or human (K562) histones are subjected to the same separation, similar profiles and retention times are observed with additional separation of primary sequence variants of H2B, H2A, and H3.
2. Regenerate the column using a linear gradient of 55–100% B over 3 min and then return to the initial conditions with a linear gradient from 100 to 25% B over 3 min.

B. *Acetic acid–methanol–$CaCl_2$ buffer system*

Equipment and reagents

- Buffer A: 20 mM $CaCl_2$, 3.5% (v/v) methanol, 0.1% (v/v) acetic acid, 25% (v/v) acetonitrile
- Buffer B: as buffer A but 70% (v/v) acetonitrile
- Zorbax SB300 C-8 (250 × 4.6 mm) reverse-phase column

Protocol 10. *Continued*

Method

1. Use the following elution conditions for the Zorbax SB300 C-8 (250 × 4.6 mm) reverse-phase column: linear gradient 0–25% B over 5 min, linear gradient 25–65% B over 30 min. The retention times of chicken erythrocyte histones under these conditions are: H5, 7 min; H1, 9 min; H2B/H4, 23 min; H2A, 26 min; and H3, 34 min.
2. Regenerate the column using a linear gradient of 65–100% B over 5 min followed by a linear return to 0% B over 5 min.

A typical elution profile using the TFA buffer system is shown in *Figure 3*.

5.3 Separation of ubiquitinated forms of calf histones H2A and H2B

A small amount of H2A (~10%) and H2B (~1.5%) exists in the post-translationally modified ubiquitinated forms uH2A and uH2B. The covalent attachment of the C-terminus of the 76 amino-acid protein ubiquitin occurs

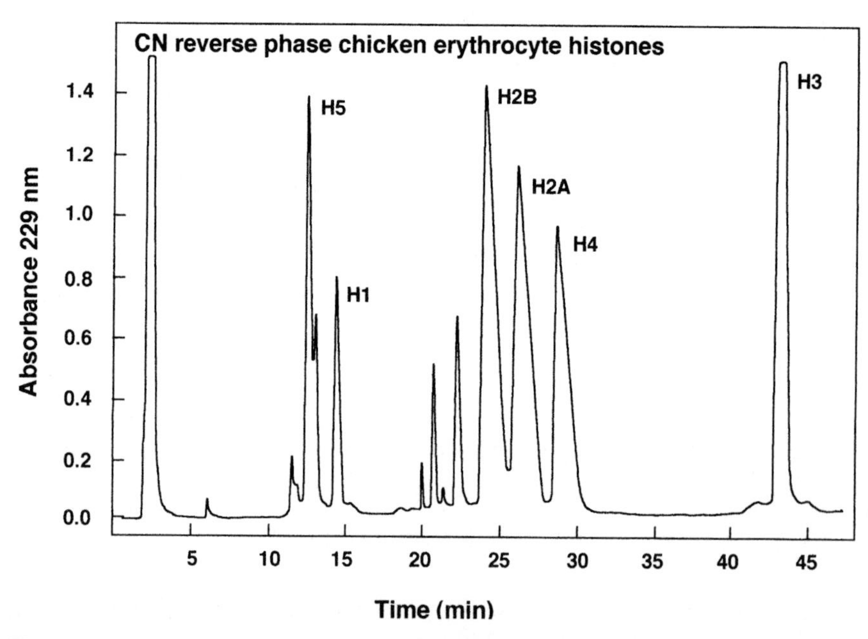

Figure 3. CN reverse-phase separation of all 6 histones from chicken erythrocytes. Column dimensions: 250 × 4.6 mm Zorbax SB300-CN. Buffer A: 0.1% (v/v) TFA; Buffer B: 0.1% (v/v) TFA in acetonitrile. Initial conditions: 25% B, flow rate 1.4 ml/min. Gradient: linear 25–32% B over 10 min; linear 32–37% B over 5 min; linear 37–42% B over 20 min; linear 42–55% B over 10 min. Column regeneration: 55–100% B over 3 min and then a linear gradient from 100–25% B over 3 min. Total loading: 1 mg.

on the side chain of lysine 119 for H2A and on lysine 120 for H2B. The method used for the separation of each of these modified species relies on the folding characteristics of the ubiquitin moiety. Under the conditions used for the above P60 fractionation the core histones exist essentially as random coils, but ubiquitin remains fully folded. Ubiquitin therefore makes little contribution to the effective size of the H2A (or H2B) molecule. Fractions from the P60 that contain uH2A (or uH2B) are separated from the unmodified H2A (or H2B) by P60 chromatography using a buffer containing 7 M urea to denature the ubiquitin moiety and thus increase the effective size of the conjugate.

Protocol 11. Fractionation of ubiquitinated histones

Equipment and reagents

- Loading buffer 1 (*Protocol 8*)
- Degassed column buffer: 7 M urea, titrated to pH 2.2 with HCl
- 200–400 mg of desalted uH2A/H2A (or uH2B/H2B) fractions from the P60 column. (It is not necessary to remove contaminating H3 from either fraction for this procedure. Note: appropriate fractions must be selected by analysis of aliquots from the P60 fractions using 15% SDS-PAGE.)
- P60 column (1400 × 27 mm) equilibrated in the above buffer. Prepare the resin by swelling and removing fines in 10 mM HCl as described earlier, and then transfer into the 7 M urea column buffer. Thoroughly degas the resin and all buffers prior to pouring and running the column.

Method

1. Dissolve protein (200–400 mg) in 4 ml of buffer 1 and leave stirring for 4 h at 4°C. Reduce the pH to 2.2 by adding concentrated HCl.
2. Elute the proteins from the column using 7 M urea, pH 2.2, at a flow rate of 3.7 ml/h at room temperature.
3. Monitor the eluant at 230 nm, collecting fractions 1/100th of the column volume.

The ubiquitinated histones elute immediately before the main H2A or H2B peak and usually co-elute with various high molecular weight proteins (and sometimes dimers of H3), see *Figure 4*. These may be removed by subsequent gel filtration on a 1400 × 15 mm Sephadex G-75 column equilibrated with 10 mM HCl (conditions under which the ubiquitin refolds).

5.4 Ion-exchange fractionation of the acetylated forms of the core histones

The acetylated forms of the calf histones H4, H3, and H2B are separated by ion-exchange chromatography. Low pressure separations are achieved using Whatman CM52 resin and HPLC separations are made using a Beckman CM-

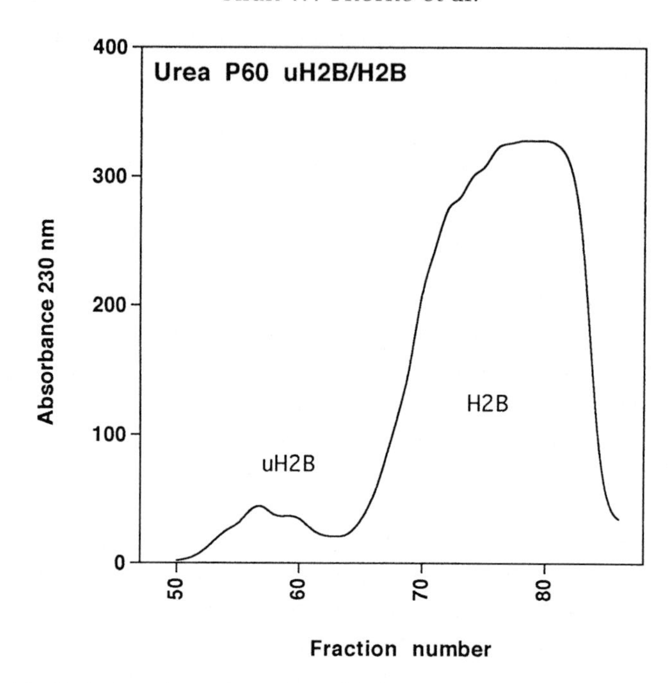

Figure 4. Bio-Rad P60 gel filtration of calf H2B in 7 M urea, pH 2.2. Column dimensions: 1440 × 26 mm. uH2B elutes at fraction 61, ahead of the expected void volume at fraction 73. Flow rate: 0.7 cm/h (3.7 ml/h).Total loading: 500 mg of H2B/uH2B.

3SW ion-exchange column, both column types being equilibrated with the same buffer. Acetylated species of each of the core histone fractions are eluted using shallow, linear salt gradients (50–300 mM NaCl) in the same buffer. Column dimensions and gradient conditions are similar for each of the core histones. For low pressure separations the use of a shallow salt gradient and a long column is essential for optimal separation of the acetylated species. Typical results are shown in *Figures 5*(*a*)–(*d*).

Protocol 12. Separation of acetylated histones

Equipment and reagents

- Column buffer: 6 M urea, 50 mM sodium acetate, pH 4.75, 0.1 mM EDTA
- Whatman CM52 (700 × 7 mm) or Beckman CM-3SW HPLC ion-exchange column (75 × 7.5 mm) equilibrated with the above buffer

Method

1. Dissolve the desalted purified histone fraction in 1–5 ml column buffer (include 10% mercaptoethanol if H3 is to be separated). Agitate gently (e.g. by rolling) for 1 h at room temperature prior to loading.

2. Elute the acetylated species with the following gradient profiles:
   (a) HPLC separations. Gradient buffers: Buffer A, column buffer; Buffer B, column buffer plus 0.3 M NaCl. Elution conditions: 5 min isocratic elution with 100% buffer A; 5 min linear gradient 0–60% buffer B; 40 min linear gradient 60–100% buffer B. Flow rate 1 ml/min throughout.
   (b) low pressure separations. Typical gradient buffers: Buffer A, column buffer plus 100 mM NaCl; Buffer B, column buffer plus 250 mM NaCl. Typical elution conditions: linear gradient buffer A to buffer B; total gradient volume 500 ml; flow rate 4.3 ml/hr.
3. Monitor the absorbance at 230 nm. Check for overlap of acetylated forms by 0.9 M acetic acid–6.25M urea PAGE (section 4) prior to pooling fractions. Dialyse exhaustively against 0.5% formic acid or desalt by means of a G25 column prior to lyophilization.

There is reduced resolution of H2B and H3 acetylated fractions compared with that achieved for the acetylated species of H4, which is probably the result of the lower fractional charge differences and the presence of some sequence variants. However, appropriate conservative cuts can provide pure acetylated H2B fractions. Acetylated species of human core histones can be achieved using the same columns. However, for H3 there is considerably reduced chromatographic resolution and overlap of the different acetylated states due to the primary sequence variants of human H3 (*Figure 5(d)*).

6. Perchloric acid (PCA) extraction of proteins from whole calf thymus cells

This procedure is very convenient for the preparation of linker histones, HMG proteins of all three types (1/2, 14/17, I/Y/I-C), and other nuclear proteins having a very high proportion of charged amino acids, e.g. thymosins, ubiquitin.

Protocol 13. PCA extraction of proteins from whole calf thymus cells

Equipment and reagents

- 5% (w/v) perchloric acid (PCA), 1 mM PMSF
- Waring blender
- Refrigerated centrifuge (Sorvall RC-3B) with (H6000 A) rotor swing out

Method

1. Freeze calf thymus directly in liquid nitrogen within minutes of animal slaughter and keep frozen at −80°C until required. Each thymus weighs approximately 200–300 g for calves that are 6–26 weeks' old. The following refers to an input of about 1.5 kg of thymus.

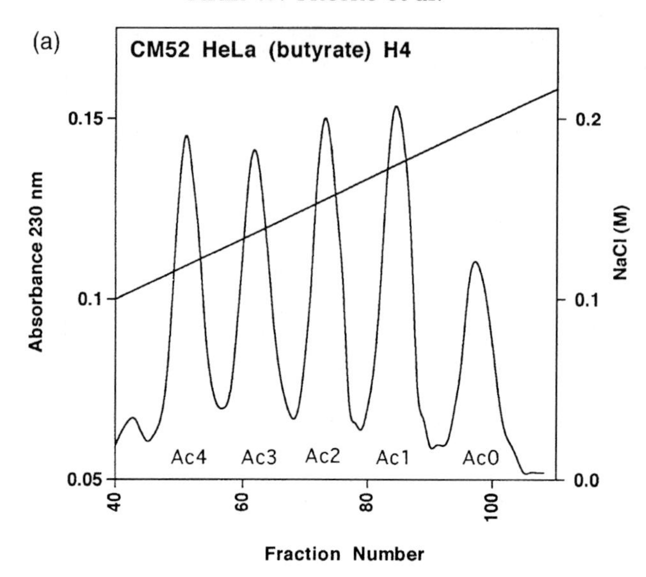

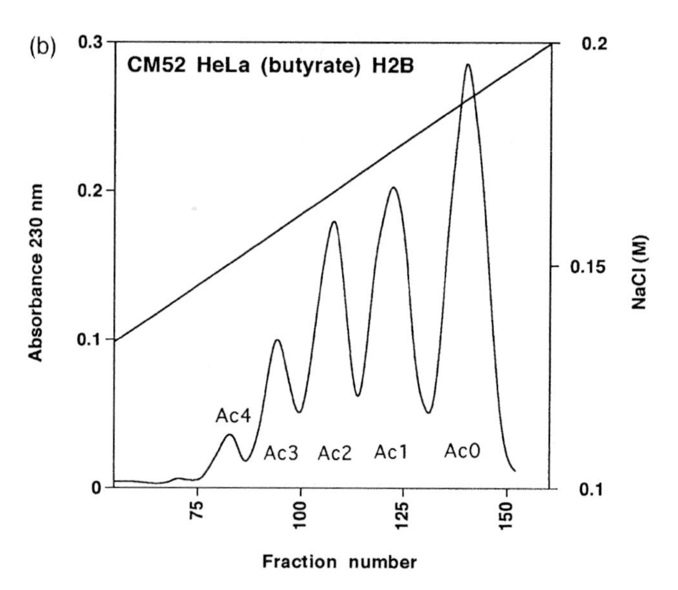

Figure 5. (a) Cation-exchange chromatography of HeLa histone H4 on Whatman carboxymethylcellulose. Column dimensions: 500 × 9 mm. Buffer: 6 M urea, 50 mM acetate, pH 4.75, 0.1 mM EDTA. Gradient: 80–250 mM NaCl over 500 ml. Flow rate: 6.8 cm/h (4.3 ml/h). Total loading: 7 mg H4. (b) Cation-exchange chromatography of HeLa histone H2B on Whatman carboxymethylcellulose. Column dimensions: 730 × 7 mm. Buffer: 6 M urea, 50 mM acetate, pH 4.75, 0.1 mM EDTA. Gradient: 100–200 mM NaCl over 320 ml. Flow rate: 6 ml/h. Total loading: 5 mg H2B. (c) HPLC cation-exchange chromatography of calf histone H3 on Beckman CM3SW. Column dimensions: 75 × 7.5 mm. Buffer A: 6 M urea, 50 mM acetate, pH 4.75, 0.1 mM EDTA; Buffer B: Buffer A plus 0.3 M NaCl. Gradient:

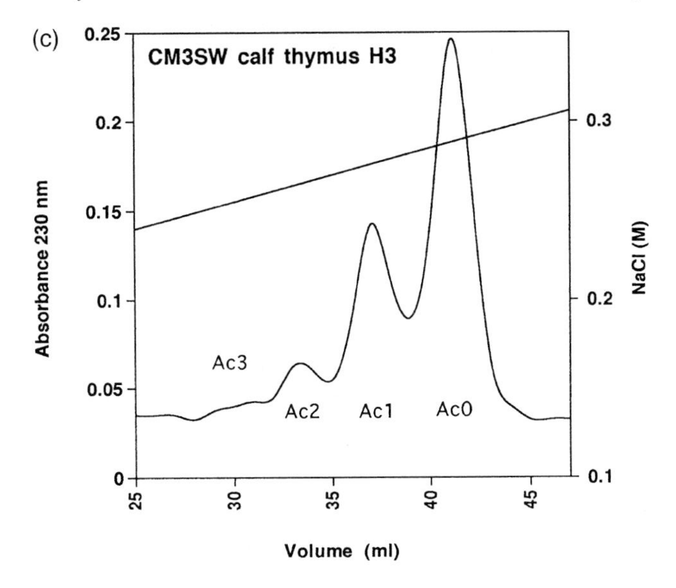

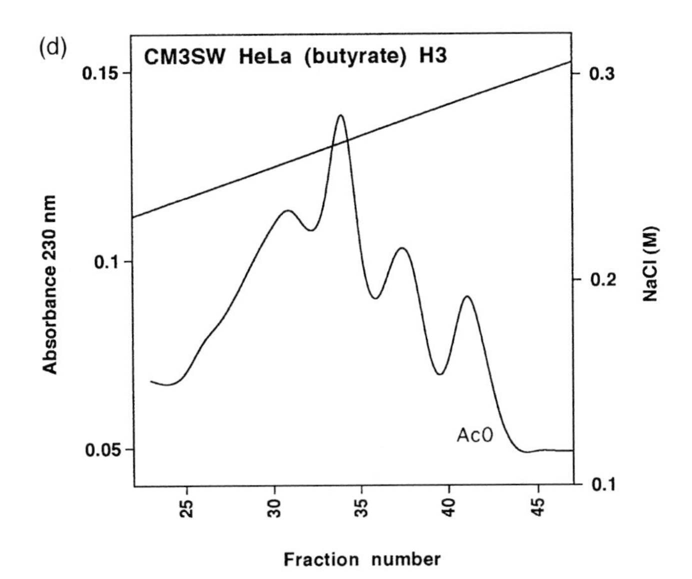

isocratic A over 5 min; linear 0–60% B over 5 min; 60–100% B over 40 min. Flow rate: 1 ml/min.Total loading: 0.5 mg H3. (d) HPLC cation-exchange chromatography of HeLa H3 on Beckman CM3SW. Column dimensions: 75 × 7.5 mm. Buffer A: 6 M urea, 50 mM acetate, pH 4.75, 0.1 mM EDTA; Buffer B: Buffer A plus 0.3 M NaCl. Gradient: isocratic A over 5 min; linear 0–60% B over 5 min; 60–100% B over 40 min. Flow rate: 1 ml/min. Total loading: 0.5 mg H3. Only the last eluted peak represents a single homogenous fraction (Ac0). The peak eluted at fraction 37 is a mixture of Ac1 and Ac0 from two different primary sequence variants, whilst the earlier eluting peaks are more complex mixtures of primary sequence variants.

Protocol 13. *Continued*

2. Suspend the frozen tissue in a 5-litre beaker containing 2.5 litres of water, 1 mM PMSF at 20 °C in the cold-room (4 °C) until the thymus is soft enough to work (30–60 min).
3. Remove and discard all fatty tissue and large blood vessels by cutting away using a pair of scissors. Cut the remaining tissue into small (2 cm) pieces (representing about two-thirds of the original weight).
4. Blend 300 g of tissue in 650 ml of 5% (w/v) PCA, 1 mM PMSF at 4 °C for 3 min using a Waring blender at full power in the cold-room.
5. Centrifuge the suspension in 1-litre buckets at 2000 *g* for 10 min. at 4 °C. Remove any surface fat, taking care to retain as much of the supernatant as possible.
6. Reblend the pellets as before in 650 ml of 5% (w/v) PCA, 1 mM PMSF and repeat the centrifugation process. Pool the two supernatants. If desired, core histones may be extracted from the pellets with 3 M NaCl or 0.5 M HCl at this stage.
7. Gradually add 3.5 volumes of acetone/concentrated HCl (400:1 (v/v)) to each volume of PCA extract and leave for 1 h at 4 °C to precipitate histone H1.
8. Centrifuge for 30 min at 2000 *g* at 4 °C. Remove and retain the supernatant containing the non-histone proteins. Wash the H1 pellet three times with dry acetone, as described for total histone, prior to vacuum drying at room temperature. Further purification of H1 can be achieved at this stage by dissolving this pellet at about 10 mg/ml in 10 mM HCl and gradually adding 3 volumes of acidified acetone. Leave for 1 h at 4 °C for full precipitation. This yields $>$ 95% histone H1.
9. Add a further 3 volumes of acetone/HCl to the supernatant retained in step 8. Leave overnight at 4 °C in 5-litre beakers covered with aluminium foil to obtain maximal precipitation of the non-histone proteins (nuclear and cytoplasmic).
10. Centrifuge for 30 min at 2000 *g* at 4 °C. Remove the supernatant. Wash and centrifuge the pellet 3 times with acetone/HCl and vacuum dry.

6.1 Purification of the high-mobility group proteins (HMGs) and thymosins

All the following steps can be carried out at room temperature. All chromatographic separations should be monitored by UV absorption and then fractions analysed by SDS-PAGE to decide which fractions to pool for the subsequent stage (see *Figure 6*).

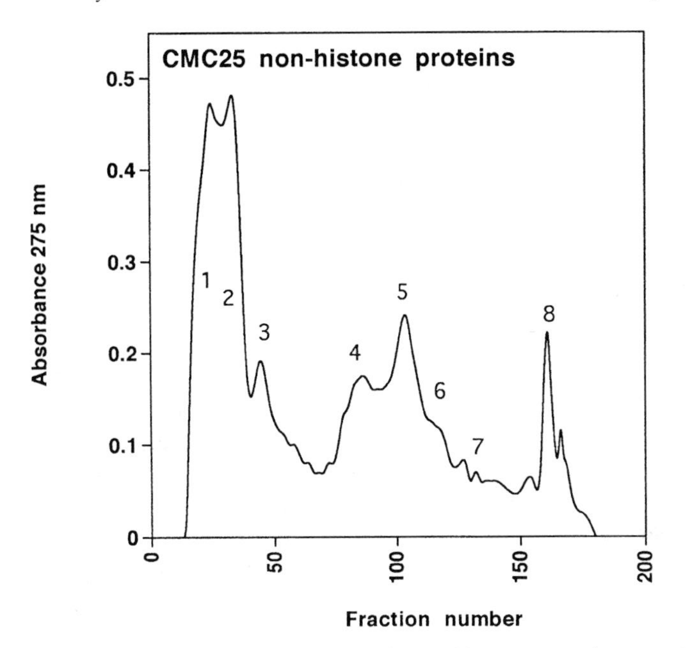

Figure 6. Cation-exchange chromatography of non-histone proteins on Sephadex CMC25. Column dimension: 300 × 20 mm. Buffer: 7.5 mM Na-borate, pH 8.8. Elution: 1 column volume (94 ml) of buffer yields peaks 1–3; 1 column volume of buffer + 0.15 M NaCl yields peak 4; 1 column volume of buffer + 0.2 M NaCl yields peak 5; linear gradient 0.2–1.0 M NaCl over 3 column volumes elutes peaks 6–8. Total loading: 4 g. Components: 1, thymosins, ubiquitin, acidic proteins, some HMG1; 2, HMG1; 3, HMG2; 4, HMG1; 5, HMG2; 6, HMG14; 7, HMG17; 8, histone H1.

Protocol 14. Purification of HMG and other nuclear proteins

Equipment and reagents

- 300 × 20 mm Sephadex CM C25 column
- 1300 × 25 mm Sephadex G75 column
- 300 × 10 mm Sephadex SP A25 ion-exchange column
- 10 mM Na-acetate buffer, pH 4.5
- 250 × 4.6 mm Bio-Rad, Hi-Pore C4 reverse-phase column
- 10 mM mercaptoethanol, 7.5 mM Na-borate, pH 8.8, with no NaCl and with 0.15 M, 0.2 M, and 1 M NaCl

Method

1. Dissolve 1–4 g of non-histone protein precipitate in 10–20 ml of water at 20 °C (requires 1–2 h with gentle agitation).
2. Remove any insolubles by centrifugation at 2000 *g* for 10 min.
3. Load the supernatant onto a 300 × 20 mm Sephadex CM C25 column equilibrated with the borate buffer (no NaCl). Elute proteins initially using one column volume of the same buffer at a flow rate of 1 ml/min, collecting 10 ml fractions. The run-through (peaks 1, 2, and 3)

Protocol 14. *Continued*

contain prothymosin-α and the β-thymosins, ubiquitin, some HMG1, and a number of acidic proteins.

4. Continue the elution using one column volume of 0.15 M NaCl in the same buffer. This elutes HMG1 (peak 4).
5. Continue elution with a further column volume of 0.2 M NaCl in the same buffer. This elutes the remaining HMG1 followed by some HMG2 (peak 5).
6. Finally, use a linear salt gradient from 0.2 to 1.0 M NaCl over three column volumes to elute HMG2, HMG14 (peak 6), HMG17, and HMGI (peak 7). Whilst this is the approximate order of elution the separation is far from complete.
7. Purify HMG14 and HMG17 by HPLC, using the reverse-phase column and a linear gradient of 0.05% (v/v) TFA/H_2O (buffer A) to 0.05% (v/v) TFA/acetonitrile (buffer B). Gradient profile: linear gradient 0–30% B over 30 min and then 30–80% B over 20 min, both with a flow rate of 0.5 ml/min. HMG17 and 14 are separated in that order during the first 30 min gradient. Contaminating HMG2 elutes at about 35% acetonitrile.
8. Further purify HMG1 and HMG2 by re-cycling down a smaller CM C25 column and applying the same salt steps and conditions as above. Lyophilize and desalt proteins on a Sephadex G15 column before loading.
9. Purify the thymosins and other proteins in the run-through peak from step 3 by gel filtration on a 1300 × 25 mm Sephadex G75 column equilibrated in 10 mM HCl, 0.02% Na-azide, flow rate 18 ml/h. Appropriate fractions (as determined by SDS-PAGE) are pooled and lyophilized.
10. Further purify the thymosins and ubiquitin on a 300 × 10 mm Sephadex SP A25 ion-exchange column equilibrated in 10 mM sodium acetate, pH 4.5 and eluted at a flow rate of 20 ml/h using the same buffer. The run-through peak contains pure prothymosin-α. Use a salt gradient of 0–0.5 M NaCl in the same buffer to elute the β-thymosins (β_4 then β_9) first and ubiquitin later.

If further purification of ubiquitin is required, this can be done by gel filtration on Sephadex G50 in 10 mM HCl (which does not denature it).Thymosins can be purified further using a C4 reverse-phase column. Approximate yields from 1.5 kg of thymus are: histone H1, 5–10 g; HMG1, 50–100 mg; HMG2, 30–60 mg; HMG14, 10–20 mg; HMG17, 20–30 mg; prothymosin-α, 5–10 mg; thymosin-β_4, 20–50 mg; ubiquitin, 20–50 mg; and core histones, 5–15 g.

References

1. Panyim, S., Bilek, D., and Chalkley, R. J. (1971). *J. Biol. Chem.*, **246**, 4206–15.
2. Simon, R. H. and Felsenfeld, G. (1979). *Nucleic Acids Res.*, **6**, 689–96.
3. Laemmli U. K. (1970). *Nature*, **227**, 680–5.
4. Panyim, S. and Chalkley, R. J. (1969). *Arch. Biochem. Biophys.*, **130**, 337–46.
5. Bonner, W. M., West, M. H. P., and Stedman, J. D. (1980). *Eur. J. Biochem.*, **109**, 17–23.
6. van der Westhuyzen, D. R., Bohm, E. L., and von Holt, C. (1974). *Biochim. Biophys. Acta*, **359**, 341–5.

3

Mapping chromatin structure in *Drosophila*

LORI L. WALLRATH, MARCI J. SWEDE, and SARAH C. R. ELGIN

1. Introduction

Most of the DNA making up a eukaryotic genome is packaged into nucleosomes. Discontinuities in the nucleosome array are typically due to the binding of non-histone chromosomal proteins. These nucleosome-free regions (about 50–200 bp in length) were first identified as sites hypersensitive to digestion with DNase I and other nucleases (reviewed in refs 1, 2). DNase I hypersensitive sites have been found 5′, 3′, and within introns of genes in organisms ranging from plants to humans (3–6). DNase I hypersensitive sites frequently map to locations of gene regulatory elements. Nucleosomes have been shown to block gene transcription when positioned over *cis*-acting regulatory elements in the promoter region of a variety of genes (reviewed in refs 7–10); this suggests that the formation of DNase I hypersensitive sites encompassing *cis*-acting elements might be a critical regulatory event providing accessibility for *trans*-acting regulatory factors. Consistent with this idea, the appearance of DNase I hypersensitive sites can be tissue specific and/or developmentally regulated (11–13).

Knowledge about the chromatin structure of a given locus can often provide information concerning the interactions of histone and non-histone chromosomal proteins that bring about appropriate gene regulation. The chromatin structure of several genes in *Drosophila* has been characterized using a variety of cleavage reagents (3, 11, 13–22). When dissecting the regulatory region of a gene, it can be useful to analyse the effects of mutations on chromatin structure and the relationship to gene expression. This type of analysis has been applied in *Drosophila* by using P-element-mediated germline transformation (23). Typically, deletions or substitutions are made within the promoter region of a gene *in vitro* and the altered gene is reintroduced into the genome using P-element-mediated germline transformation. Expression and chromatin structure of the transgene are compared to that of a control transgene (with a wild-type regulatory region) or to that of the endogenous gene (24–26). For example, the effect of altered chromatin structure on heat-shock induced expression has been analysed for the *hsp26* gene of *Drosophila melanogaster* (25, 26). Transgenes also can be useful for studying alterations in chromatin structure associated with particular chromosomal environments (27–29).

This chapter focuses on chromatin structure analysis of *Drosophila* genes, transgenes, or any genomic sequence for which there is a cloned DNA fragment available to use as a probe. A protocol for the isolation of nuclei from a relatively small number of embryos, larvae, and adults is described. Protocols for the treatment of nuclei with DNase I, micrococcal nuclease (MNase), and restriction enzymes are included along with a description of the advantages and limitations of each enzyme. A discussion of how one detects and evaluates the patterns of cleavage to derive a picture of the protein–DNA interactions for a given genomic region is presented.

2. *Drosophila* culture

The choice of the developmental stage and tissue source for isolation of nuclei is an important consideration, depending on the genomic region of interest. Changes in chromatin structure generally accompany transcriptional activation of a gene. When looking at the chromatin structure of such a gene it is therefore important to isolate nuclei from an appropriate tissue at a stage when the gene is likely to be uniformly active or inactive. For housekeeping genes and heat-shock genes, which are expressed in all tissues of the organism, it is more convenient to use the entire organism for analysis (3, 14, 19, 26). This is also the case when examining the chromatin structure of a non-transcribed region, such as a simple repetitive sequence (15). When a gene is tissue and/or developmentally regulated, nuclei must be prepared from tissue isolated in bulk or by hand dissection (11, 21, 22). Methods for the mass isolation of organs from third instar larvae of *Drosophila* are available (30, 31).

2.1 Collection of embryos for nuclear preparations

Population cages (containing approximately 50 000 adult flies) can be used for collections of large numbers of *Drosophila* embryos. A thorough description of the cage construction, systematic maintenance of large *Drosophila* cultures, and embryo collection is presented elsewhere (32). Population cages are required when large-scale isolations of nuclei from carefully timed embryos are needed. For the small-scale isolation of nuclei (starting material of 1 g of embryos), it is possible to obtain sufficient embryos from adults reared in standard *Drosophila* culture bottles. This is particularly beneficial when dealing with a large number of different transformed *Drosophila* lines. Embryos are a good source of nuclei because they contain relatively low levels of degradative enzymes, but for isolation of nuclei on a small scale it is easier to collect larvae and adults as starting material.

For small-scale embryo collection, six or more bottles (6-oz. (150 ml) disPo bottles; Baxter Healthcare Corp.) with cornmeal media are filled with approximately 300 healthy adult flies each and are maintained at 25 °C. After 4 days the adults are removed. Once the adults of the next generation begin to emerge they are collected each day and placed in new bottles with a large supplement of fresh yeast paste (a thick slurry of active yeast and water). These adults are allowed to mature for 5–6 days, at which time a few thousand are placed in a small egg-collecting chamber (33) and allowed to lay eggs on a 'spoon' containing grape-juice agar media (17% grape-juice concentrate, 4% agar in water). A small paintbrush moistened with water can be used to collect the embryos from the collecting spoon. Embryos can be used fresh, or frozen in liquid nitrogen and stored at − 70 °C for later use.

2.2 Collection of larvae for nuclear preparations

Four bottles with approximately 800–1000 adult flies (5 days post-emergence) each are maintained at 25 °C (preferably 70% humidity). The adult flies are transferred to fresh bottles every 4 days. These same adults can be used to set up at least three sets of four bottles for larval collection. The larvae of the next generation should reach the third instar stage approximately 4 days after removal of the adults. When the walls of the bottle are coated with crawling larvae, the larvae can be collected with a spatula by scraping the walls of the bottle. The larvae are immediately dropped into a beaker containing liquid nitrogen. Frozen larvae can be weighed using a pre-chilled weigh-boat and stored at − 70 °C. Collections can be made for several days from the same bottles. Typically, each bottle yields 3 g or more of larvae.

2.3 Collection of adults for nuclei

Live adult flies are collected from bottles after anaesthetizing them with CO_2, and placed in a disposable screw-cap tube. These adults are frozen by immersing the tube in liquid nitrogen and can be weighed on a pre-chilled weigh-boat. Approximately 5–7 ml of adult flies weighs 1 g. Frozen adult flies are stored at − 70 °C.

3. Isolation of nuclei

The overall strategy for chromatin structure analysis is shown in *Figure 1A*. The following procedure has been used to isolate nuclei from 1 g of embryos, larvae, or adults. Embryos, larvae, and adult flies can be used fresh, or frozen in liquid nitrogen and stored at − 70 °C several years prior to use. No difference has been found between fresh and frozen material when performing low-resolution (agarose gel) chromatin structure analysis on the *hsp26* gene of *Drosophila*.

Gloves should be worn during buffer preparation and throughout the isola-

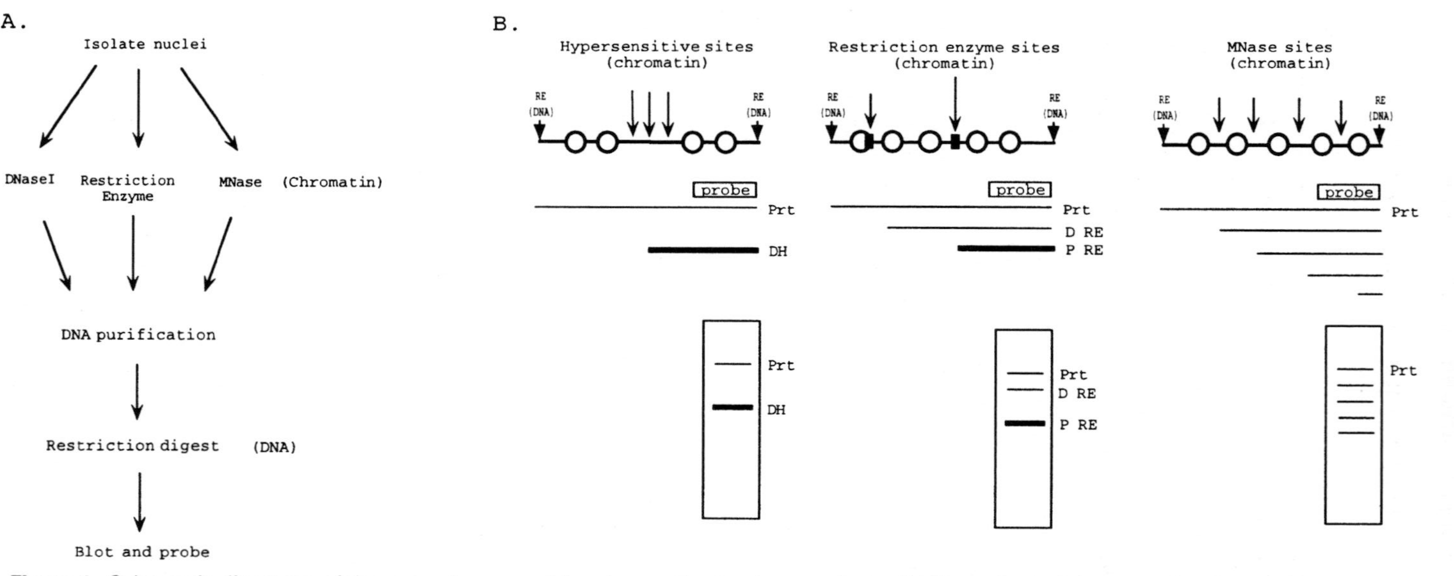

Figure 1. Schematic diagrams of the procedures used for chromatin structure analysis. (A) Flow chart of the steps for chromatin structure analysis and indirect end-labelling. Nuclei isolated from embryos, larvae, or adults are treated with a limited amount of a general chromatin cleavage reagent (DNase I or MNase) or an excess amount of a restriction enzyme. The DNA is purified and digested to completion with a second restriction enzyme for indirect end-labelling analysis. (B) Schematic diagram of digestion products resulting from treatment of nuclei with cleavage reagents and subsequent indirect end-labelling analysis. The top line represents chromatin, with DNA represented by a line and nucleosomes represented by circles. The sites preferentially digested by the indicated cleavage reagents are marked with arrows. The arrowheads represent the site of cleavage by a restriction enzyme after the DNA has been purified. The resulting products are diagrammed below with the fragment of DNA used as a probe shown. The parental restriction fragment, resulting from no chromatin cleavage within the region of interest, is labelled Prt. One of the fragments resulting from cleavage at hypersensitive sites is labelled DH. The DNA fragments resulting from restriction enzyme cleavage of the DNA at the sites proximal and distal relative to the probe are labelled P RE and D RE, respectively. A schematic diagram of an autoradiograph of a low-resolution (agarose gel) analysis of the cleavage products after case above hybridization with the probe indicated is shown for each case above.

tion of nuclei procedure to avoid contamination with the degradative enzymes present on hands. All solutions and glassware are kept on ice (4 °C) throughout the procedure. Buffer A, without spermine, spermidine, and DTT (*Protocol 1*), can be made up as a 10 × stock and stored at 4 °C. Just before use, the appropriate amount of spermine, spermidine, and DTT (from a 0.1 M stock of each stored at − 20 °C) is added to 10 × buffer A, and the solution brought to the desired volume with water to make a 1 × complete buffer A. Complete buffer A can then be used to make buffer A^+ with the addition of the appropriate amount of Nonidet P-40 and additional EDTA (*Protocol 1*). Complete buffer A can also be used to make buffer AS with the addition of the appropriate amount of sucrose (*Protocol 1*).

Spermine and spermidine have been included in the buffers used here to prevent the swelling and rupture of the nuclear membrane that can occur because of the low divalent-cation concentrations. Divalent cations have been minimized to reduce activity of endogenous nucleases. In some cases, however, the presence of polyamines is unacceptable; this depends on the type of chromatin structure analysis to be performed. Polyamines, such as spermine and spermidine, have been shown to selectively extract certain chromosomal proteins, such as HMG proteins, and can affect chromatin structure (34, 35). Generally, inclusion of polyamines is not a problem for low-resolution (agarose gel) chromatin structure analyses; it appears that the polyamines do not cause perturbations of the nucleosome array. However, a problem might arise when performing high-resolution (acrylamide sequencing gel) structural analyses. In particular, footprints of the TATA binding protein and certain transcription factors are easily lost in the presence of polyamines (G. H. Thomas and S. C. R. Elgin, unpublished observation, 1988; 36). If spermine and spermidine are left out of the buffers, it may be necessary to include divalent cations or to reduce the concentration of chelators to obtain stable nuclei. Control samples with no added enzyme should always be generated to assess the extent of endogenous nuclease activity.

Protocol 1. Isolation of nuclei from *Drosophila* embryos, larvae, and adults

Equipment and reagents

- Mortar and pestle
- 7 ml ground-glass homogenizer (Corning Science Products)
- 15 ml Dounce homogenizer with type B pestle (Wheaton Science Products)
- 7 ml Dounce homogenizer with type A and B pestles (Wheaton Science Products)
- (Calbiochem, cat. no. 475855)
- Centrifuge and swinging-bucket rotor (e.g. Sorvall HB-4 rotor, Du Pont)
- 1 × Buffer A: 60 mM KCl, 15 mM NaCl, 1 mM EDTA, 0.1 mM EGTA, 15 mM Tris–HCl, pH 7.4, 0.15 mM spermine, 0.5 mM spermidine, 0.5 mM DTT
- Buffer A+: 60 mM KCl, 15 mM NaCl, 13 mM EDTA, 0.1 mM EGTA, 15 mM Tris–HCl, pH 7.4, 0.15 mM spermine, 0.5 mM spermidine, 0.5 mM DTT, 0.5% Nonidet P-40 (Sigma)
- Buffer AS: 60 mM KCl, 15 mM NaCl, 1 mM EDTA, 0.1 M EGTA, 15 mM Tris–HCl, pH 7.4, 0.15 mM spermine, 0.5 mM spermidine, 0.5 mM DTT, 0.3 M sucrose

Protocol 1. *Continued*

Method

1. Pulverize 1 g of embryos, larvae, or adults in liquid nitrogen using a pre-chilled mortar and pestle.
2. After the liquid nitrogen has evaporated, transfer the homogenate to a beaker containing 3 ml of buffer A^+ and mix. The frozen ground material should be completely saturated with buffer A^+. If it is not, add more buffer.
3. Transfer the suspension to a 7 ml ground-glass homogenizer and disrupt with 5 strokes of the pestle.
4. Transfer the homogenate to a 15 ml Dounce homogenizer, disrupt with 10 strokes using a pestle (type B), and filter through one layer of Miracloth.
5. Rinse the ground-glass homogenizer and the Dounce homogenizer with 3 ml of buffer A^+ and filter the suspension through the same layer of Miracloth. Squeeze the mesh (or cloth) with gloved fingers to complete filtration.
6. Pipette the filtrate onto 1 ml of buffer AS and centrifuge in a swinging bucket rotor at 3000 r.p.m. (1500 *g*) at 4 °C for 5 min.
7. Resuspend the pellet in 3 ml of buffer A^+, transfer the nuclear suspension to a 7 ml Dounce homogenizer, and disperse with 5 strokes of the type A pestle. Pipette the suspension onto 1 ml of buffer AS, and centrifuge as for step 6.
8. Resuspend the pellet in 3 ml of buffer A, transfer the suspension to the 7 ml Dounce homogenizer, and disperse with 5 strokes of the type B pestle.
9. Centrifuge at 2000 r.p.m. (700 *g*) for 5 min at 4°C, and resuspend the pellet in the appropriate amount of a particular digestion buffer (see *Protocols 2, 3,* and *4*).

4. Treatment of nuclei with cleavage reagents

4.1 DNase I treatment

DNase I binds across the minor groove of the double helix and cleaves the phosphodiester backbone. Nucleosome-free regions of DNA are hypersensitive to this cleavage. Limited digestion with DNase I and low-resolution (agarose gel) analysis can be used to qualitatively assess the sensitivity of a genomic region (20). When used in conjunction with indirect end-labelling (see section 6.2), such digestion can be used to map DNase I hypersensitive sites (3, 11, 13, 19). DNase I can also be used as a footprinting reagent. High-resolution (acrylamide sequencing gel) analysis of DNA, derived from nuclei treated with DNase I and subsequently cleaved by restriction enzyme digestion for indirect end-labelling, will show protected sites at locations in which particular nucleotides are likely to be contacted by protein factors. When DNA is associated with a histone octamer, a 10–11 bp interval cleavage pattern spanning the region associated with the core histones is detected by high-resolution analysis (19, 37).

Several separate digestions with increasing amounts of DNase I are generally carried out to achieve a wide size range of digestion products. We recommend using increasing amounts of nuclease for a fixed time, rather than keeping the concentration of nuclease constant and increasing the incubation time; this results in samples which all have the same time of exposure to endogenous nucleases. A sample of nuclei should be incubated without nuclease to check the level of activity of endogenous nucleases. It is absolutely essential to include a 'naked' (protein-free) DNA control when mapping chromatin structure. DNase I has been shown to cleave preferentially in DNA sequences that exhibit certain structural features (e.g. preferred helical twist; 38) and specific sequences (consensus sequence 5′ A/T Y A/T Y V N 3′, where V is not T, and N is either A, C, G, or T; 39). The protein-free DNA should be treated with DNase I to a similar extent of digestion as the DNA in the nuclei. A bona fide DNase I hypersensitive site would be one present or enhanced in the nuclear sample relative to the protein-free DNA sample at equivalent levels of digestion. In practice, this means that the average size of the DNA, as judged by gel electrophoresis, should be approximately the same.

Protocol 2. DNase I treatment of nuclei

Reagents

- DNase I digestion buffer: 60 mM KCl, 15 mM NaCl, 15 mM Tris–HCl, pH 7.4, 0.25 M sucrose, 3 mM $MgCl_2$, and freshly added DTT to 0.5 mM
- DNase I (20 000 units/ml in 10 nM Tris-HCl, pH 7.5, 50 mM NaCl, 10 mM $MgCl_2$ 1 mM DTT; 50% glycerol; Boehringer Mannheim Biochemicals cat. no. 776785)

Method

1. Resuspend the pellet of isolated nuclei (containing approximately 200 μg of genomic DNA) (*Protocol 1*, step 9) in 1 ml of DNase I digestion buffer.
2. Incubate 250 μl aliquots with 0, 1, 2, or 4 μl of DNase I on ice (4 °C) for 3 min with occasional gentle agitation. Determine the actual concentration of DNase I to use empirically, as this will vary with each lot of DNase I and the manufacturer.
3. Add 5 μl of 0.4 M EDTA (pH 8.0) to each sample to terminate the reaction and proceed to *Protocol 5*.

4.2 Restriction enzyme treatment

Restriction enzymes are used to quantitatively assess the accessibility of given sites within isolated nuclei. This assay, using defined sites, allows one to generate a limit digest; this is not possible with less specific cleavage reagents, such as DNase I and MNase, which give only qualitative results. Restriction enzyme sites within DNase I hypersensitive, nucleosome-free regions generally show 50–80% cleavage in isolated nuclei; in contrast, sites within nucleosome-associated DNA show 6–8% cleavage (25, 26, 40, 41). Intermediate values have been obtained for sites at the edge of a nucleosome (41; L.L. Wallrath, unpublished results).

It is extremely important to have an excess amount of restriction enzyme present when determining the quantitative accessibility of a site. To ensure restriction enzyme excess, a titration experiment is performed in which identical amounts of nuclei are treated with increasing units of restriction enzyme for an equal amount of time. For experimental purposes, 10% more than the minimal number of units of a restriction enzyme that gives the maximum amount of cleavage is the concentration to use. A titration experiment must be carried out for each restriction enzyme used. Typically, we find that 200 to 400 units of a particular restriction enzyme are sufficient for maximal cleavage within nuclei in a suspension containing approximately 25 μg of genomic DNA.

Protocol 3. Treatment of nuclei with restriction enzymes

Reagents

- 0.1 M PMSF in isopropanol (Sigma)
- 10 mg/ml leupeptin in dH_2O (Sigma)
- 100 units/ml aprotinin in dH_2O (Sigma)
- 1 × restriction enzyme buffer (manufacturer's specifications)

Method

1. Resuspend the pellet of nuclei (containing approximately 200 μg of genomic DNA) (*Protocol 1*, step 9) in 1.47 ml of the appropriate restriction enzyme digestion buffer.
2. Add protease inhibitors PMSF, leupeptin, and aprotinin to final concentrations of 0.5 mM, 5 μg/ml, and 0.1 unit/ml, respectively (these are required because of the long incubation time used).
3. Add an excess amount of the particular restriction enzyme to 250 μl aliquots of the resuspended nuclei. Determine the amount of restriction enzyme to use empirically, such that the addition of increasing units of enzyme does not result in more cleavage at a given site.
4. Mix the sample by pipetting up and down several times.

5. Incubate the digest at 37 °C (or manufacturer's specification) for 45 min with occasional agitation.
6. Add 5 μl of 0.4 M EDTA (pH 8.0) to each sample to terminate the reaction and proceed to *Protocol 5*.

4.3 Micrococcal nuclease (MNase) treatment

MNase digestion provides information about the nucleosome organization over a region of DNA. MNase preferentially cleaves within the linker DNA, between histone octamers, and is inhibited from cleavage at locations in which the DNA is associated with histone octamers. DNA purified from MNase-cleaved chromatin can be used directly for analysis to determine if a particular cloned region is packaged in a regular array of nucleosomes, or it can be digested with a restriction enzyme for subsequent indirect end-labelling experiments to map the regions sensitive to MNase digestion (see below).

Once again, we stress the importance of including a naked DNA control. MNase shows a very pronounced sequence preference for a general endonuclease; preferential cleavage is seen in regions of alternating A and T residues (for consensus sequences see ref. 42).

Protocol 4. Treatment of nuclei with MNase[a]

Reagents

- MNase digestion buffer: 60 mM KCl, 15 mM NaCl, 15 mM Tris–HCl, pH 7.4, 0.25 M sucrose, 1 mM $CaCl_2$, and freshly added DTT to 0.5 mM
- 0.08 units/μl MNase (Sigma cat. no. N5386) in 5 mM Tris–HCl, pH 7.5, 25 μM $CaCl_2$

Method

1. Resuspend the pellet of nuclei (containing approximately 200 μg of genomic DNA) (*Protocol 1*, step 9) in 1 ml of MNase digestion buffer.
2. Add 0, 2, 4, and 6 μl of MNase to 250 μl aliquots of resuspended nuclei. Determine the amount of MNase empirically, as it will vary with each lot of MNase and the manufacturer.
3. Incubate at 25 °C for 2 min with occasional gentle agitation.
4. Add 5 μl of 0.4 M EDTA (pH 8.0) to terminate each reaction and proceed to *Protocol 5*.

[a] See also *Protocol 3*, Chapter 9.

5. DNA isolation

5.1 Isolation of DNA from nuclei

Protocol 5. Purification of DNA from nuclei

Equipment and reagents

- 10 mg/ml proteinase K in dH_2O (Sigma cat. no. P2308)
- Sarcosyl lysis buffer: 50 mM Tris–HCl, pH 8.0, 100 mM EDTA, 0.5% (w/v) sodium laurylsarcosine
- RNase A digestion buffer: 50 mM Tris–HCl, pH 8.0, 10 mM EDTA, 100 mM NaCl
- 10 mg/ml RNase A (Sigma cat. no. R6513) in 10 mM Tris–HCl, pH 7.5, 15 mM NaCl
- TE buffer: 10 mM Tris–HCl, pH 8.0, 1 mM EDTA
- Phenol:chloroform:isoamyl alcohol (24:12:1)
- Chloroform:isoamyl alcohol (24:1)
- Water-saturated ethyl ether
- Microcentrifuge

Method

1. Collect nuclei from the the terminated reaction mixture (step 3, 6, or 4 of *Protocols 2, 3,* or *4,* respectively) by centrifugation for 30 sec in a microcentrifuge.
2. Resuspend the pellet in 250 μl of sarcosyl lysis buffer.
3. Add 3 μl of proteinase K and incubate at 37 °C overnight.
4. Add 83 μl of 2 M NaCl.
5. Add 400 μl of phenol:chloroform:isoamyl alcohol (25:24:1), vortex for 5 min, and centrifuge for 5 min in a microcentrifuge at room temperature.
6. Remove the aqueous phase and place it in a new tube. Optional: Check the extent of digestion by loading 5 μl of the supernatant on an agarose/TAE mini gel[a], and determine the DNA sizes of the separated fragments by electrophoresis and staining the gel with ethidium bromide.
7. Add 400 μl of isopropanol to the aqueous phase, mix by inversion and place at −20 °C for 20 min to overnight, or on dry ice for 10 min.
8. Centrifuge in a microcentrifuge for 10 min at 4 °C.
9. Decant the isopropanol and add 200 μl of 70% (v/v) ethanol in distilled H_2O.
10. Repeat the centrifugation as in step 8.
11. Decant the 70% ethanol and vacuum dry the pellet.
12. Resuspend the pellet in 120 μl of RNase A digestion buffer.
13. Add 3 μl of RNase A and incubate for 1 h at 37 °C.
14. Add 40 μl of 2 M NaCl.

15. Repeat the extraction as in step 5 above, twice for restriction enzyme-treated nuclei and once for DNase I- and MNase-treated nuclei.
16. Add 400 μl of chloroform:isoamyl alcohol (24:1) to the aqueous phase, vortex for 5 min, centrifuge in a microcentrifuge for 5 min at room temperature, and place the aqueous phase in a new tube.
17. Add 1 ml of water-saturated ethyl ether to the aqueous phase.
18. Vortex for 5 min, centrifuge in a microcentrifuge for 5 min at room temperature, and pipette off and discard the organic phase.
19. Place tubes with their lids open at 65°C to eliminate residual ether (approximately 5 min).
20. Add 400 μl of 100% ethanol and precipitate at −20°C overnight or for 10 min on dry ice.
21. Centrifuge in a microcentrifuge for 10 min at 4°C.
22. Decant the ethanol and add 200 μl of 70% ethanol.
23. Centrifuge in a microcentrifuge for 10 min at 4°C.
24. Vacuum dry the pellet and resuspend the DNA in 20 μl of TE buffer.

[a] See ref. 43.

We can suggest a few trouble-shooting measures that may need to be taken when performing the steps in *Protocol 5*. First, a gelatinous liquid layer, instead of a firm pellet, is occasionally found following the centrifugation in step 8. This may be due to incomplete digestion of proteins by the proteinase K treatment. If the gelatinous layer is present, pipette the top isopropanol layer off using a micropipettor, rather than decanting. Add the 70% ethanol and invert the tube to mix the layers. At this point a white precipitate should become apparent. Proceed with the centrifugation in step 10. Second, it is sometimes difficult to resuspend the DNA in TE buffer after step 24; this is particularly true for the control samples that were not treated with any cleavage reagent. If the purified DNA is going to be used for restriction enzyme digestion it is important that it be completely in solution to avoid partial digestion products. There are several measures used to aid in resuspending the DNA in TE buffer. We recommend heating the sample at 65°C for 10 minutes, vortexing, as well as mechanically disrupting the pellet by pipetting the sample up and down in a pipette tip. Overnight storage at 4°C, rather than at −20°C, prior to restriction enzyme digestion increases the likelihood of resuspending the DNA. Lastly, because of the many manipulations in the DNA isolation procedure, it is likely that the concentration of DNA in each tube will vary (see *Figure 2*, lane 2 of DNase I digestion series as an example). If subsequent manipulations require very similar DNA concentrations it is best to visualize a small aliquot (1 μl of purified DNA from step 24) by electrophoretic-size separation on an agarose/TAE mini-gel stained with ethidium bromide to estimate relative concentrations among samples. Alternatively, the DNA concentration can be determined by UV spectrophotometry (43).

DNA isolated from nuclei that have not been treated with cleavage reagents can be used as a source for 'naked' DNA controls. However, for most purposes, genomic DNA from approximately 200 adult flies is a convenient source, since larger quantities of DNA can be obtained per gram of organism.

5.2 Isolation of DNA from adult flies

The following is a protocol for isolating genomic DNA from adult flies which is modified slightly from that published by Bender *et al.* (44).

Protocol 6. Isolation of DNA from adult flies

Equipment and reagents

- 7 ml ground-glass homogenizer (*Protocol 1*)
- Grinding buffer: 0.1 M NaCl, 0.2 M sucrose, 0.1 M Tris–HCl, 0.05 M EDTA, pH 9.1, freshly added sodium dodecyl sulfate to 0.5% and diethyl pyrocarbonate to 1%
- Elutip-d column and elution buffer recipe (Schleicher and Schuell)
- RNase A buffer (*Protocol 5*)
- RNase A (*Protocol 5*)
- Centrifuge and rotor with adaptors to hold 15 ml glass tubes (e.g. Sorval SS 34 rotor, Du Pont)

Method

1. Collect approximately 100–200 adult flies by anaesthetizing with CO_2.
2. Transfer the flies to a 7 ml ground-glass homogenizer containing 2 ml of grinding buffer and grind quickly at room temperature.
3. Transfer the homogenate to a 15 ml thick-walled glass tube and incubate at 65 °C for 30 min.
4. Add 0.3 ml of 8 M potassium acetate and incubate on ice (4 °C) for 30 min.
5. Centrifuge at 9100 r.p.m. (10 000 *g*) for 5 min at 4 °C.
6. Mix the supernatant with an equal volume of ethanol.
7. Let stand at room temperature for 5 min.
8. Centrifuge at 9100 r.p.m. (10 000 *g*) for 5 min at room temperature.
9. Rinse the pellet with 80% (v/v) ethanol in dH_2O.
10. Dry the pellet under vacuum being extremely careful not to under- or overdry.
11. Resuspend the pellet in 1 ml of low-salt buffer for purification through an Elutip-d column. The DNA solution may appear cloudy due to residual carbohydrates and eye pigments.

12. Purify the DNA through the Elutip-d column according to the manufacturer's specification (eliminate the pre-filter).
13. After ethanol precipitation, perform steps 21–24 inclusive of *Protocol 5* and resuspend the pellet in 100 μl of RNase A buffer.
14. Add 4 μl of RNase A and incubate at 37 °C for 1–3 hours.
15. Add sodium acetate to 0.2 M and then add 250 μl of 100% ethanol.
16. Precipitate the DNA on ice for 10 min or at – 20 °C overnight.
17. Perform steps 21–24 inclusive of *Protocol 5* and resuspend the pellet in 50 μl of TE buffer.

NB The yield from 200 flies is about 60 μg of DNA.

6. Chromatin structure analysis

6.1 Nuclease sensitive sites in bulk chromatin

To examine the general sensitivity of a genomic region, the DNase I cleavage products can be examined immediately following DNA isolation. Of the DNA sample (*Protocol 5*, step 24), one-half (10 μl) is loaded directly onto a 1% agarose/TAE gel (43) (*Figure 2*). For optimizing mapping results (see below) there should be a good distribution in the size of cleavage products for the different amounts of DNase I used (*Figure 2*). After size separation by gel electrophoresis, the DNA is transferred to a membrane and hybridized with a fragment of DNA from the cloned region of interest labelled with α-^{32}P using random primers (43, 45). These DNA samples can also be used to generate quantitative data by looking at the relative rate of disappearance of mapped restriction fragments (46).

DNA isolated from MNase-treated nuclei can be analysed directly to determine whether a given cloned region is packaged into a regular nucleosome array. The DNA (10 μl from *Protocol 5*, step 24) is separated by size on a 1.2–1.5% agarose/TAE gel; the electrophoresis is carried out overnight, at 4°C to minimize diffusion of the fragments. A typical nucleosome ladder should be visible upon staining with ethidium bromide (*Figure 2*). The DNA is transferred to a membrane and probed with an α-^{32}P-labelled fragment from the region of interest. If the region is packaged into nucleosomes, a ladder of hybridization products with approximately 200 bp intervals should appear. However, if the region of interest is not packaged in such an array, a smear comprising heterogeneous sized fragments will be produced after hybridization and detection (for example, see ref. 28, *Figure 5*).

6.2 Nuclease sensitive sites mapped by indirect end-labelling (see also Chapter 9, *Protocol 7*)

The overall scheme for chromatin structure analysis using indirect end-labelling is shown in *Figure 1B*. Indirect end-labelling (47, 48) involves cleavage of the

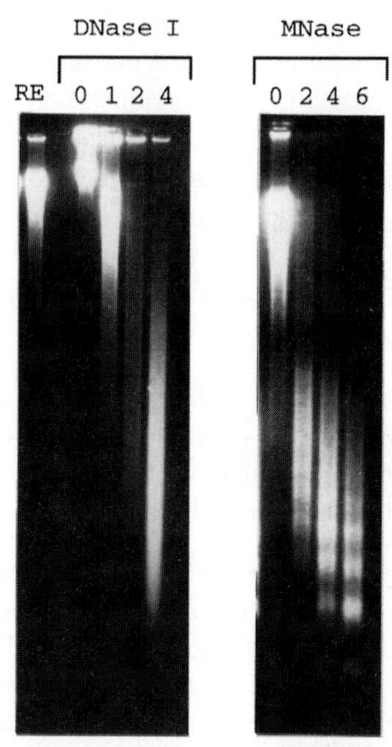

Figure 2. Ethidium bromide stained agarose/TAE gel of the DNA cleavage products isolated from nuclei treated with a restriction enzyme, DNase I, or MNase. DNA isolated from nuclei treated with increasing amounts of DNase I or MNase is shown. The number of microlitres of the digestive reagent used is shown above each lane (see *Protocols 2* and *3* for details). DNA from nuclei treated with a restriction enzyme (RE) is also shown (see *Protocol 4* for details).

purified DNA at restriction sites bracketing the region of interest (*Figure 1B*) (see Chapter 9, *Protocol 7*). The DNA is separated by size on an agarose gel (low-resolution analysis) or on a sequencing type acrylamide gel (high-resolution analysis). For the protocols provided here, we typically use the entire sample from a restriction enzyme treatment of nuclei (20 μl from step 24, *Protocol 5*) and half of the sample (10 μl) from a DNase I or MNase digest for low-resolution analysis. The probe (typically 400 bp to 1 kb) is chosen so that it abuts either end of the fragment released by restriction enzyme digestion (*Figure 1B*). This probe DNA is labelled with α-^{32}P for radioactive detection or with digoxigenin for non-radioactive detection (Boehringer Mannheim Biochemicals) using random primers (45). Non-radioactive detection can be used for assessing the accessibility of a site with restriction enzymes, but is not currently sensitive enough for the detection of patterns resulting from DNase I and MNase digestion (L. Wallrath, unpublished results). Probes can also be labelled single-stranded DNA or RNA (19, 49).

Indirect end-labelling methodology can be used to assign the location of DNase I hypersensitive sites and to determine their relative accessibility (*Figure 3*). Restriction enzyme digestion in combination with indirect end-

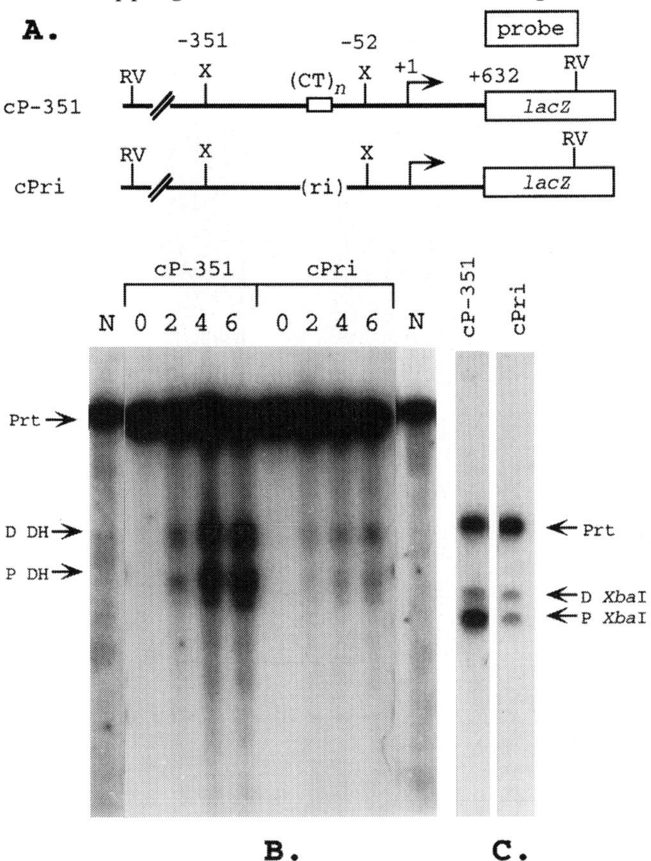

Figure 3. Mapping the DNase I hypersensitive sites and quantitating the accessibility of the proximal *Xba*I site upstream of *hsp26* transgenes. (A) Diagram of *hsp26* transgenes (not to scale). Transgene cP-351 (50) contains *hsp26* sequences from –351 to +632, with respect to the transcription start site (+ 1), fused to the *E. coli lacZ* gene. $(CT)_n$ represents a stretch of alternating C and T residues which are thought to bind the GAGA transcription factor (25). X represents the *Xba*I cleavage sites found within the regulatory heat-shock elements. RV indicates the *Eco*RV restriction enzyme sites used for indirect end-labelling. Transgene cPri is identical to cP-351 but has the $(CT)_n$ region substituted by a random insert (ri) of salmon sperm DNA (50). (B) Autoradiograph of the DNA isolated from nuclei of larvae treated with DNAse I, purified and cleaved with *Eco*RV. Naked DNA partially digested with DNase I and then completely digested with *Eco*RV is shown in the lanes labelled N. The probe used was *lacZ* sequences abutting the *Eco*RV site. Prt represents the band resulting from no cleavage within the *hsp26* upstream region in nuclei. P DH and D DH denote products resulting from cleavage at the proximal and distal DNase I hypersensitive sites upstream of *hsp26*, respectively. Note the relative reduction in accessibility of the DNase I hypersensitive sites of the cPri transgene. (C) Autoradiograph of the DNA isolated from nuclei treated with *Xba*I, purified and cleaved with *Eco*RV. The probe used was the *lacZ* sequence. Prt represents the fragment resulting from no digestion at the *Xba*I sites in nuclei. P *Xba*I and D *Xba*I represent the fragments arising from digestion at the proximal and distal *Xba*I sites, respectively. Scanning of the autoradiograph with a densitometer showed that the proximal *Xba*I site was 52% accessible in cP-351 and 8% accessible in cPri, showing that the $(CT)_n$ element plays a major role in establishing these DNase I hypersensitive sites.

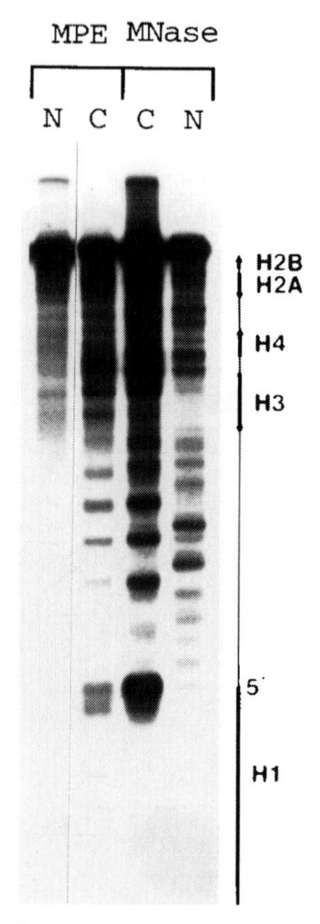

Figure 4. Mapping the MPE•Fe(II) and MNase cleavage sites along the histone gene region. Autoradiograph of DNA from nuclei isolated from embryos and treated with MNase or MPE•Fe(II), purified and cleaved with *Bgl*II for indirect end-labelling (lanes labelled C). The DNA was separated by size on a 1.6% agarose gel, transferred to a membrane, and hybridized with an α-^{32}P-labelled *Bgl*II–*Bam*HI fragment of plasmid B5 (see ref. 15 and references therein). Naked DNA that was treated with MPE•Fe(II) or MNase are in lanes labelled N. The chromatin patterns are in agreement, while the DNA patterns differ. This indicates that the reagents differ in their preferred cleavage sites on DNA, but are sensitive to chromatin packaging, cleaving preferentially in the linker between core histones. Note the reduction in sequence specificity of MPE•Fe(II) in the naked DNA control lane. A diagram of the histone genes (arrows) are shown to the right of the autoradiograph.

labelling has been used to determine the accessibility of a restriction site within a DNase I hypersensitive site and to map the boundaries of DNase I hypersensitive sites (*Figure 3*) (25, 26, 41, 51). DNA (20 μl from *Protocol 5*, step 24) from restriction enzyme-treated nuclei is separated by size on an agarose/TAE gel, transferred to a membrane, and probed with a fragment of DNA that abuts the endpoint obtained by restriction digestion of the purified DNA (*Figure 1B*). Either radioactive or non-radioactive methodology can be used for detection (see above). The autoradiograph or lumigraph can be scanned by a densitometer for quantitative measurements of the intensity of the bands. Cleavage at the site most proximal to the probe (if there is more than one) can be quantitated by determining the intensity of the band generated from cleavage at the most proximal site and dividing by the sum of the intensities of all the bands within a given lane. When nuclei are prepared from hand-dissected tissue, the amount of DNA may be insufficient for conventional blot and probe technology. LMPCR can be used to amplify the DNA purified from restriction enzyme-treated nuclei isolated from dissected tissue (see Chapter 4 and ref. 22). For LMPCR analysis, primers are used for amplification (and labelling) of the cleavage products prior to size separation on agarose or acrylamide gels. Additional information on LMPCR is presented in Chapter 4.

MNase digestion in conjunction with indirect end-labelling analysis can be used to map the locations of regions relatively protected from MNase digestion. Protection from cleavage suggests, but does not prove, that the region is packaged as a nucleosome. Additional independent evidence is needed to demonstrate that protection comes from the presence of a nucleosome. This can come from high-resolution DNase I analysis; a 10–11 bp periodic cleavage should be evident along the length of the nucleosome (19, 37). Lack of digestion at a restriction site within the suspected nucleosomal region has served as supportive evidence in several instances (41, 51, 52). If cleavage with MPE•Fe(II) generates a pattern similar to that obtained with MNase, this further supports the hypothesis of a nucleosome (53). MPE•Fe(II) cleaves with greatly reduced sequence specificity compared to MNase, simplifying mapping studies (*Figure 4*) (15). This chemical has recently become commercially available from Sigma Chemical Co. Protocols for MPE•Fe(II) cleavage are available elsewhere (15, 54). Lastly, nucleosomal boundaries have been mapped by primer extension analysis of DNA from core particles released from chromatin by extensive digestion with MNase (Chapter 8 and refs 55, 56).

Acknowledgements

We thank Iain Cartwright and Quinn Lu for allowing us to modify and present figures from previous publications. We thank members of the Elgin lab. for a critical reading of the manuscript. Work carried out in this laboratory is supported by grants GM31532 and HD23844 from the National Institutes of Health to S. C. R. Elgin. L. L. Wallrath was supported by NIH National Service Award GM14516.

References

1. Elgin, S.C.R. (1988). *J. Biol. Chem.*, **263**, 19259.
2. Gross, D.S. and Garrard, W.T. (1988). *Annu. Rev. Biochem.*, **57**, 159.

3. Wu, C. (1980). *Nature*, **286**, 854.
4. Gwynn, B., Lyford, K.A., and Birkenmeier, E.H. (1990). *Mol. Cell. Biol.*,**10**, 5244.
5. Brignon, P., Lepetit, M., Gigot, C., and Chaubet (1993). *Plant Mol. Biol.*, **22**, 1007.
6. Miyashita, T., Yamamoto, H., Takenoto, Y., Nozaki, M., Morita, T., and Matsushiro, A. (1993). *Gene*, **125**, 151.
7. Grunstein, M. (1990). *Annu. Rev. Cell. Biol.*, **6**, 643.
8. Hayes, J.J. and Wolffe, A.P. (1992). *BioEssays*, **14**, 597.
9. Workman, J.L. and Buchman, A.R. (1993). *Trends in Biochem.*, **18**, 90.
10. Becker, P.B. (1994). *BioEssays*, **16**, 541.
11. Shermoen, A.W. and Beckendorf, S.K. (1982). *Cell*, **29**, 601.
12. Fronk, F., Tank, G.A., and Langmore, J.P. (1990). *Nucl. Acids Res.*, **18**, 5255.
13. Cartwright, I.L. (1987). *EMBO J.*, **6**, 3097.
14. Wong, Y.C., O'Connell, P., Rosbash, M., and Elgin, S.C.R. (1981). *Nucl. Acids Res.*, **9**, 6749.
15. Cartwright, I.L., Hertzerg, R.P., Dervan, P.B., and Elgin, S.C.R. (1983). *Proc. Natl Acad. Sci.*, **80**, 3213.
16. Eissenberg, J.C., Kimbrell, D.A., Fristrom, J.W., and Elgin, S.C.R. (1984). *Nucl. Acids Res.*, **12**, 9125.
17. Cartwright, I.L. and Elgin, S.C.R. (1986). *Mol. Cell. Biol.*, **6**, 779.
18. Ramain, P., Bourouis, M., Dretzen, G., Richards, G., Sovbkowiak, A., and Bellard, M. (1986). *Cell*, **45**, 545.
19. Thomas, G.H. and Elgin, S.C.R. (1988). *EMBO J.*, **7**, 2191.
20. Hayashi, S., Ruddell, A., Sinclair, D., and Grigliatti, T. (1990). *Chromosoma*, **99**, 391.
21. Jackson, J.R. and Benyajati, C. (1992). *Nucl. Acids Res.*, **20**, 5413.
22. Schloβherr, J., Eggert, H., Paro, R., and Cremer, S. (1994). *Mol. Gen. Genet.*, **243**, 453.
23. Rubin, G.M. and Spradling, A.C. (1982). *Science*, **218**, 348.
24. Ramain, P., Giangrande, A., Richards, G., and Bellard, M. (1988). *Proc. Natl Acad. Sci. USA*, **85**, 2718.
25. Lu, Q., Wallrath, L.L., Allan, B.D., Glaser, R.L., Lis, J.T., and Elgin, S.C.R. (1992). *J. Mol. Biol.*, **225**, 985.
26. Lu, Q., Wallrath, L.L., Granok, H., and Elgin, S.C.R. (1993). *Mol. Cell. Biol.*, **13**, 2802.
27. Weising, K., Bohn, H., and Kahl, G. (1990). *Dev. Genet.*, **11**, 233.
28. Allshire, R.C., Javerzat, J.-P., Redhead, N.J., and Cranston, G. (1994). *Cell*, **76**, 157.
29. Wallrath, L.L. and Elgin, S.C.R. (1995). *Genes Dev.*, **9**, 1263.
30. Jowett, T. (1986). In *Drosophila: a practical approach* (ed. D.B. Roberts), pp. 275–86. IRL Press, Oxford.
31. Ashburner, M. (ed.) (1989). In *Drosophila: a laboratory manual.* pp. 280–89. Cold Spring Harbor Laboratory Press, Cold Spring Harbor, NY.
32. Shaffer, C.D., Wuller, J.M., and Elgin, S.C.R. (1994). In *Methods in cell biology* (ed. L.S.B. Goldstein and E.A. Fyrberg), Vol. 44, pp. 99–108. Academic Press, New York.
33. Santamaria, P. (1986). In *Drosophila: a practical approach* (ed. D.B. Roberts), pp. 275–286. IRL Press, Oxford.
34. Wagner, C.W., Hamana, K., and Elgin, S.C.R. (1992). *Mol. Cell. Biol.*, **12**, 1915.

35. Van den Broeck, D., Van der Straeten, D., Van Montagu, M., and Caplan, A. (1994). *Plant Physiol.*, **106**, 559.
36. Pfeifer, G.P. and Riggs, A.D. (1991). *Genes Dev.*, **5**, 1102.
37. Jackson, J.R. and Benyajati, C. (1993). *Nucl. Acids Res.*, **21**, 957.
38. Lomonossoff, G.P., Butler, P.J.G., and Klug, A. (1981). *J. Mol. Biol.*, **149**, 745.
39. Herrera, J.E. and Chaires, J.B. (1994). *J. Mol. Biol.*, **236**, 405.
40. Jack, R.S., Moritz, P., and Cremer, S. (1991). *Eur. J. Biochem.*, **202**, 441.
41. Verdin, E., Paras, P. Jr., and Van Lint, C. (1993). *EMBO J.*, **12**, 3249.
42. Flick, J.T., Eissenberg, J.C., and Elgin, S.C.R. (1986). *J. Mol. Biol.*, **190**, 619.
43. Sambrook, J., Fritch, E.F., and Maniatis, T. (ed.) (1989). In *Molecular cloning, a laboratory manual* (2nd edn), pp. 6.3–6.20, B.20, B.23. Cold Spring Harbor Laboratory Press, NY.
44. Bender, W., Spierer, P., and Hogness, D.S. (1983). *J. Mol. Biol.*, **168**, 17.
45. Feinberg, A.P. and Vogelstein, B. (1984). *Anal. Biochem.*, **137**, 266.
46. Jantzen, K., Fritton, H.P., and Igo-Kemenes, T. (1986). *Nucl. Acids Res.*, **15**, 6085.
47. Wu, C., Bingham, P.M., Livak, K.J., Holmgren, R., Elgin, S.C.R. (1979). *Cell*, **16**, 797.
48. Nedospasov, S.A. and Georgiev, G.P. (1980). *Biochem. Biophys. Res. Comm.*, **29**, 532.
49. Saluz, H.P. and Jost, J.P. (ed.) (1987). In *A laboratory guide to genomic sequencing*, pp. 111–24. Birkhauser Verlag, Basel.
50. Glaser, R.L., Thomas, G.H., Siegfried, E.S., Elgin, S.C.R., and Lis, J.T. (1990). *J. Mol. Biol.*, **211**, 751.
51. Straka, S. and Hörz, W. (1991). *EMBO J.*, **10**, 361.
52. Archer, T.K., Cordingly, M.G., Wolford, R.G., and Hager, G.L. (1991). *Mol. Cell. Biol.*, **11**, 688.
53. Richard-Foy, H. and Hager, G.L. (1987). *EMBO J.*, **6**, 2321.
54. Lu, Q., Wallrath, L.L., and Elgin, S.C.R. (1993). In *Methods in molecular genetics* (ed. K.W. Adolf), Vol. 1, pp. 333–57. Academic Press, San Diego, CA.
55. Bresnick, E.H., Rories, C., and Hager, G.L. (1992). *Nucl. Acids Res.*, **20**, 865.
56. Georgel, P., Dretzen, G., Jagla, K., Bellard, F., Dubrovsky, E., Calo, V., and Bellard, M. (1993). *J. Mol. Biol.*, **234**, 319.

4

In vivo footprint and chromatin analysis by LMPCR

ARTHUR D. RIGGS, JUDITH SINGER-SAM, and GERD P. PFEIFER

1. Introduction

In vitro footprinting, as introduced by Galas and Schmitz (1), has become a common procedure which uses cloned DNA and purified DNA binding proteins or nuclear extracts. *In vivo* footprinting, performed by genomic sequencing after treatment of nuclei or intact cells with dimethyl sulfate (DMS), was introduced by Church *et al.* (2, 3) and improved by Becker *et al.* (4) and Saluz and Jost (5). However, for mammalian single-copy genes, the technique required the use of high levels of radioactivity and remained difficult until the advent of ligation-mediated PCR (LMPCR) (6, 7). LMPCR takes advantage of the sensitivity of PCR and the specificity of nested PCR. This enables a high-resolution, nucleotide-level analysis of chromatin, and only a microgram or less of total genomic DNA is needed. LMPCR was introduced by Mueller and Wold (6) for DMS footprinting and extended by Pfeifer and his colleagues for use in DNA sequencing (7), methylation analyses (7, 8), DNase I footprinting (9, 10), nucleosome analyses (9), and UV photofootprinting (11, 12). In essence, LMPCR is a procedure for detecting the presence of rare breaks in the phosphodiester backbone of DNA. Generally, the breaks or nicks are the result of enzyme or chemical agents that directly, or indirectly after additional treatment, provide a 5′ phosphate.

As schematically outlined in *Figure 1*, a key step of the LMPCR procedure is primer extension which starts from a target site specified by a gene-specific oligonucleotide primer and continues until the 5′ end of the template molecule is reached. This step creates blunt ends on all template molecules that participate in primer extension and have a 5′ end within the range of primer extension. The next step is blunt-end ligation of a common oligonucleotide linker onto the primer-extended molecules that have a blunt end and a 5′ phosphate. After the ligation step, all participating molecules have a defined sequence on both the left and right side and thus can serve as templates for exponential PCR amplification. Because all sizes of template molecules

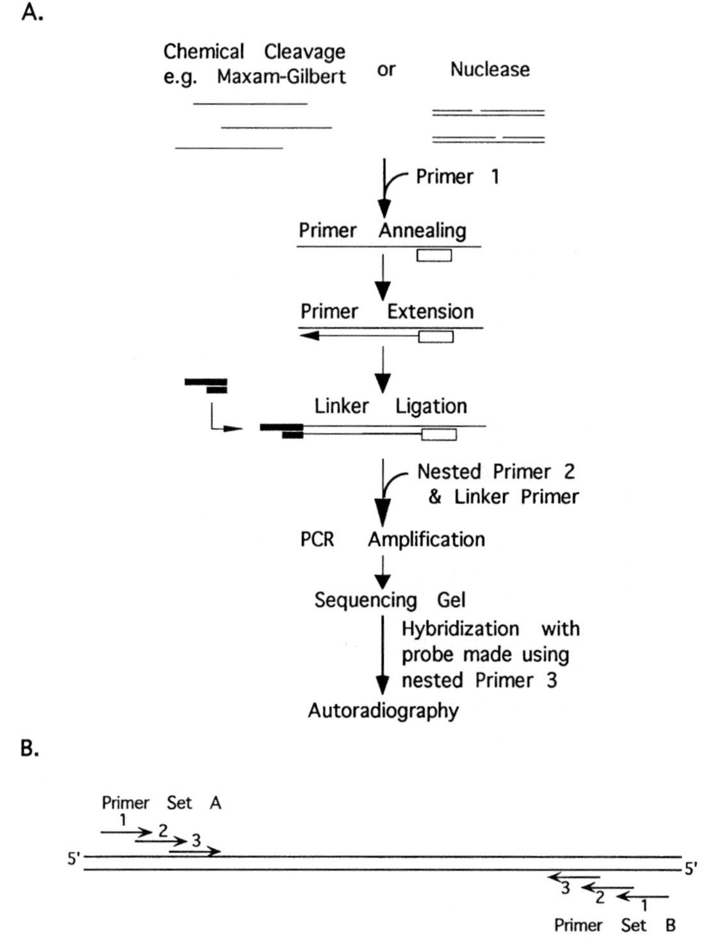

Figure 1. (A) Outline of the LMPCR method. (B) Arrangement of primers to analyse both strands of a 200–300 bp region. Primer set A will reveal footprints on the upper strand, while primer set B will reveal footprints on the lower strand. Primer 1 of each set is used for a primer extension reaction. Nested primer 2 is used for PCR, and nested primer 3 is used to make a hybridization probe using a cloned fragment or PCR product as a template.

Figure 2. LMPCR autoradiograms. The left panel shows the results of an *in vivo* footprinting experiment using dimethyl sulfate (DMS) to investigate changes in transcription factors induced by exposure of the cells to UV light (+UV). Sites of altered DMS reactivity are indicated by circles. Some sites have increased (closed circles) or decreased (open circles) reactivity compared to the Maxam–Gilbert controls (Lanes 1–4, labelled C, C+T, A+G, and G). The right panel shows the results of an *in vivo* DNase I footprinting experiment. Lane 13, labelled 'DNA' is a control in which purified DNA is treated with DNase I. Lanes labelled 'cells' were permeabilized cells treated with DNase I. The cells were either stimulated (+UV) or unstimulated (–UV) by treatment with UV light. Footprint regions of increased cleavage relative to the naked DNA control are indicated by closed rectangles; protected regions are indicted by open rectangles. Adapted from Rozek and Pfeifer (10).

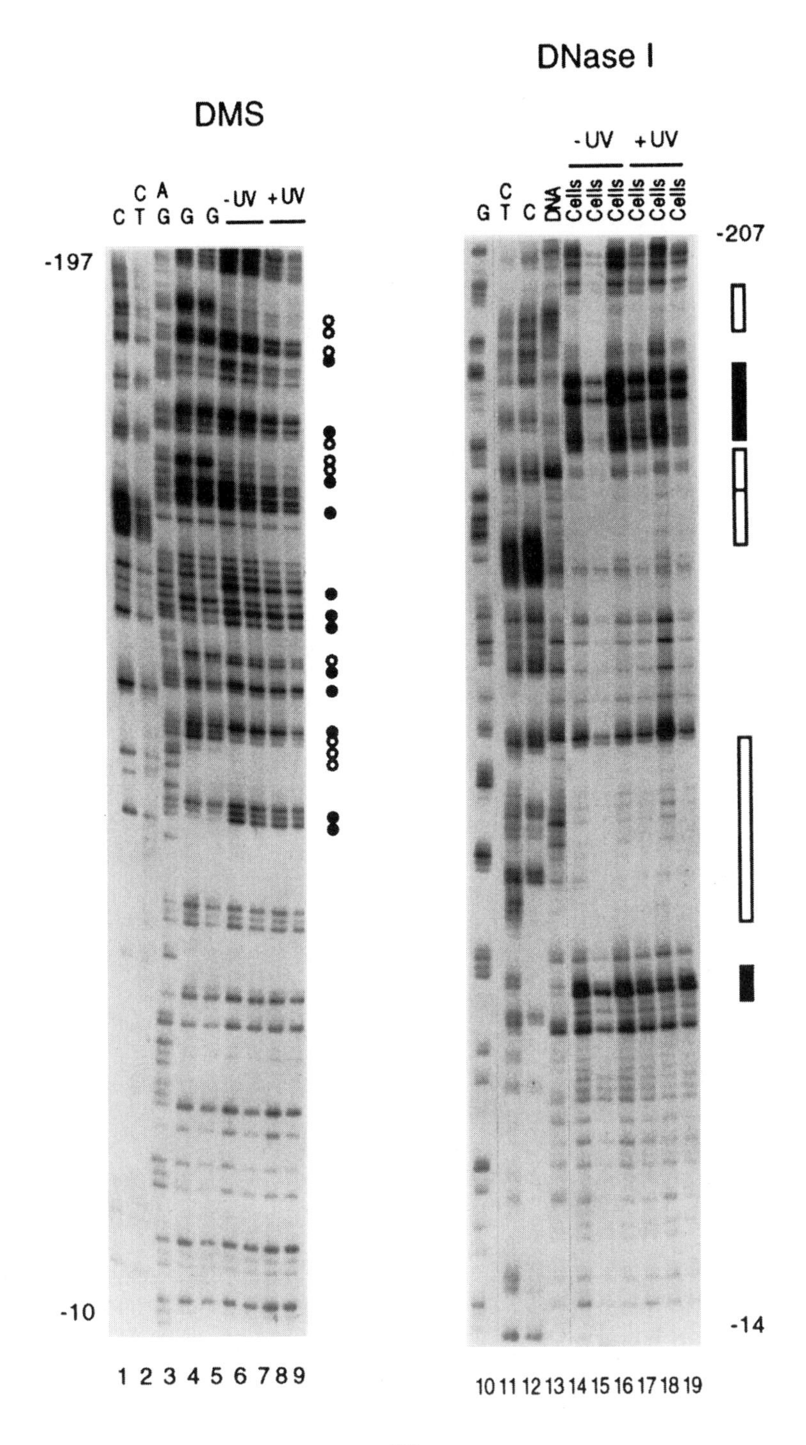
DMS
DNase I
C
T
C
A
G
G
G
-UV
+UV
G
C
T
C
DNA
Cells
Cells
Cells
Cells
Cells
Cells
-UV
+UV
-197
-10
-207
-14
1 2 3 4 5 6 7 8 9
10 11 12 13 14 15 16 17 18 19

have the same primer target sequences, PCR amplification is usually of similar efficiency, so after gel electrophoresis and hybridization an autoradiogram is obtained looking much like a DNA sequencing ladder (*Figure 2*). LMPCR can, in fact, be used to determine the base sequence if the initial cleavage is by Maxam–Gilbert base-specific chemical agents (7, 13). Maxam–Gilbert lanes are usually included as controls to confirm a sequence and to determine the precise nucleotide-level location of any detected feature such as an alkylation or nuclease footprint.

Over 60 *in vivo* genomic footprinting studies have now been published (reviewed in ref. 14). These *in vivo* studies mostly have confirmed *in vitro* studies with nuclear extracts, but there have been numerous, interesting exceptions. Confirming observations first made by Becker *et al.* (4), several studies have found that protein factors present in the cell and nuclear extracts are not bound to DNA *in vivo*, as evidenced by the lack of a footprint (6, 9, 15–19). For these and other reasons, chromatin accessibility and 'open' and 'closed' states of chromatin have come to be recognized as important, often dominant aspects of somatic gene and chromosome function (20, 21). Recently Schlossherr *et al.* (22) used the Steigerwald *et al.* (23) adaptation of LMPCR to analyse the accessibility of restriction enzymes to *Drosophila* chromatin; they found that gene silencing by position-effect variegation does not involve major changes in chromatin accessibility. This type of accessibility experiment should be increasingly performed as part of future chromatin structure–function studies.

LMPCR is a rather robust technique if performed with care, but steps prior to LMPCR are also important, as are controls and methods for visualizing the LMPCR-generated ladder of bands. For this reason, procedures for these pre- and post-LMPCR steps are included as part of this chapter. Taken as a whole, the procedure takes several days, so attention to detail is highly recommended. The procedure is still relatively young and many steps have not yet been fully optimized, but since there are a number of variables, changes should be made cautiously.

2. Preparation of genomic DNA

DNA used for LMPCR should be of high quality. Some DNA preparations give poor quality ladders and/or show missing bands (24). The cause of poor results has usually not been determined, but the DNA isolation procedure given in *Protocol 1*, which includes nuclei preparation, has given consistently good results. In this procedure, nuclei are isolated according to the method of Wijnholds *et al.* (25), and DNA is isolated essentially as described by Saluz and Jost (5) for genomic sequencing before the invention of LMPCR. Many other DNA isolation procedures are very likely to be adequate, but, again, one should be careful about the quality of the DNA. A recent paper by Glasel (26) is of interest in this regard, because DNA preparations that seem adequately pure by absorbance measurements may be heavily contaminated with proteins. If nuclei are not isolated, an extra step to remove RNA (see *Protocol 2B*, step 9) is recommended because RNA fragments are good primers for DNA polymerases. *Protocol 1* will give very high molecular weight DNA. For some experiments this may be desired, but usually the high viscosity and resulting slowness in dissolving the DNA is an inconvenience which can be eliminated by treatment with a restriction enzyme that does not cut within the region of LMPCR analysis.

Protocol 1. Preparation of nuclei and DNA from tissue culture cells

Reagents

- PBS: 137 mM NaCl, 2.7 mM KCl, 1.5 mM KH_2PO_4, 8.0 mM Na_2HPO_4, pH 7.4 (Irvine Scientific or Gibco-BRL)
- Buffer A: 0.3 M sucrose, 60 mM KCl, 15 mM NaCl, 60 mM Tris–HCl, pH 8, 0.5 mM spermidine, 0.15 mM spermine, 2 mM EDTA
- Buffer B: 150 mM NaCl, 5 mM EDTA, pH 7.8
- Buffer C: 20 mM Tris–HCl, pH 8, 20 mM NaCl, 20 mM EDTA, 1% SDS, and 600 μg/ml proteinase K (weighed out and added just before use)
- TE buffer: 10 mM Tris–HCl, pH 8.0, 1 mM EDTA. 1/10 TE: Ten-fold dilution of TE
- RNase A: 100 mg/ml in 10 mM Tris–HCl, pH 7.5 (Boehringer Mannheim; DNase free, or prepared according to ref. 27)
- 0.1 M Tris–HCl, pH 8
- Nonidet P-40 (Sigma)
- Reagents and solutions for phenol/chloroform extraction[a]: chloroform, liquid phenol (equilibrate phenol with 0.1 M Tris–HCl, pH 8.0)
- Reagents and solutions for ethanol precipitation[a]: ethanol, 3 M sodium acetate, pH 5.2, 7 M ammonium acetate, 0.5 M EDTA, pH 7.7
- Clinical Centrifuge (IEC)

Method

1. Wash 10^7 to 10^8 cells with 30 ml of PBS at 37 °C and resuspend in 4 ml of buffer A at 5 °C.
2. Add 1 volume of cold buffer A containing 1% (v/v) Nonidet P-40 and incubate for 5 min on ice. Check to see that the cells have lysed and the nuclei are intact.
3. Pellet nuclei by centrifugation at 5 °C for 5 min at 1000 *g* in a clinical centrifuge.
4. Wash nuclei by resuspension in 15 ml of cold buffer A and recentrifugation.
5. Resuspend nuclei in 5 ml of cold buffer B.
6. Add 1 volume of buffer C and incubate for 3 h at 37 °C.
7. Digest RNA by adding RNase A to 100 μg/ml and incubating for 1 h at 37 °C.
8. Extract using phenol–chloroform. First add 0.5 volume of phenol saturated with 0.1 M Tris–HCl, pH 8, and mix by rocking or inverting the tube for 4–5 min; then add 0.5 volume chloroform and spin briefly to separate the phases. Remove the aqueous phase using a wide-

Protocol 1. *Continued*

mouth pipette, and repeat the phenol–chloroform extraction 1 to 2 times more, until there is little material at the interface.

9. Extract once with chloroform to reduce the amount of phenol in the aqueous phase.
10. Ethanol-precipitate by adding one-tenth volume of 3 M sodium acetate, pH 5.2, and 2.5 volumes of room temperature ethanol. Mix by inverting, and after 10 min at room temperature collect the DNA either by a brief centrifugation or by collecting on a glass rod or pipette (spooling)[b]. Wash the DNA pellet with 70% ethanol and air-dry.
11. Dissolve the pellet in TE buffer (pH 8) or 1/10 TE buffer using 200 μl per 10^7 cells. This solution is kept as a stock DNA solution. Unless rare breaks are being studied, the solution is kept frozen at −20 °C. To avoid even rare DNA breakage during freezing, keep the solution at 4 °C, but it is advisable to add a drop of chloroform to prevent mould growth.
12. To reduce viscosity and to enable more rapid solubilization after ethanol precipitation, digest the DNA used for LMPCR with a restriction enzyme which does not cut within the region sequenced, using an appropriate buffer, and then purify the DNA by phenol–chloroform extraction and ethanol precipitation.
13. Dissolve the DNA pellet obtained after step 12 in water or 1/10 TE buffer and adjust the concentration to 1–10 μg/μl. If the DNA is to be used for chemical cleavage, resuspend it in water. If the DNA is to be used for the determination of rare breaks, resuspend it in 1/10 TE buffer to better avoid even rare strand breaks.

[a] Chemicals should be reagent grade or the highest purity available. The source is not known to be critical unless specified otherwise. Solutions and procedures very commonly performed in molecular biology are as in Sambrook *et al.* (27) unless specified otherwise.
[b] If nuclei are not prepared as, for example, in the DNase I footprint procedure (*Protocol 2B*), the preparation will contain considerable amounts of RNA, therefore spooling is recommended, as well as the additional RNase treatment described in *Protocol 2B*, step 9.

3. Treatment of intact cells with dimethyl sulfate or DNase I for *in vivo* footprinting

Treatment of cells with the alkylating agent dimethyl sulfate (DMS) is the easiest procedure to perform (*Protocol 2A*). DMS readily penetrates the cell membrane and alkylates DNA, mostly at the 7 position of guanine. After isolation, the DNA can be readily cleaved at alkylated sites by treatment with piperidine. Conveniently, piperidine treatment provides single-stranded DNA molecules with the necessary 5′ phosphate. The DMS method (6) has proven to be very useful for the identification of many transcription factors (14), but has the disadvantage that not all protein–DNA contacts are revealed. Nucleosomes, for example, do not affect the DMS reactivity pattern and thus are transparent to this assay (28).

Numerous *in vitro* footprinting studies have established that DNase I is a very useful footprinting agent, since it is sensitive to most nucleoprotein complexes as well as DNA structural variations such as width of the minor groove (29, 30). DNase I has often been used for chromatin structure studies on isolated nuclei, but we prefer to use intact cells and a cell permeabilization method (31, 32). The isolation of nuclei often results in less clear footprints or the occasional complete absence of transcription factor footprints (9). Spermidine and spermine are included in the most commonly used nuclei isolation buffers to stabilize nuclei, but this also seems to destabilize transcription factors at the human PGK promoter (9). The lysolecithin-based cell permeabilization procedure given in *Protocol 2B* was adapted from Zhang and Gralla (32) and works well for cells that grow firmly attached as a monolayer. For other cells that grow less firmly attached or in suspension, we have had good results treating cells in suspension with lysolecithin and washing by centrifugation. Nonidet P-40 can also be used in place of lysolecithin, as described by Hornstra and Yang (33). For each cell line it is essential to determine the optimum concentration of lysolecithin or Nonidet P-40 experimentally, ascertaining by microscopic examination that the cells remain intact. Importantly, both the concentration of the permeabilization agent and the concentration of nuclease to use must be experimentally determined by assaying for the frequency of introduced nicks.

For most initial LMPCR experiments, we recommend the use of DNA that has a weight-average, single-strand size of 300–600 nt as estimated by comparison with DNA size standards on alkaline agarose gels. It should be noted that estimation of ethidium bromide-stained DNA by visual comparison with size standards gives a weight-average size, but signal strength of an LMPCR band depends on the number of molecules, which will be proportional to the number-average size. For a Gaussian distribution of fragment sizes, the number-average size will be about one-half the weight-average size (34). For most experiments, the difference in number- and weight-average size will not be of great significance, but this aspect should be kept in mind. We routinely determine both double- and single-strand length by neutral and alkaline agarose electrophoresis, respectively, as described by Sambrook *et al.* (27). Some chromatin structural features of the inactive X chromosome are lost rapidly after *in vivo* DNA nicking (L. Wang and A. D. Riggs, unpublished), so after initial experiments have established that a good LMPCR ladder can be obtained, one may want to minimize the nick density as well as the time of DNase treatment prior to DNA extraction, in order to observe chromatin in its least perturbed state.

Although we have so far only used cell permeabilization procedures for DNase I or micrococcal nuclease (9), the procedure should be general and, for example, could be used with restriction enzymes. For enzymes or chemical procedures that do not leave a 5′ phosphate, one must add a kination step, as for micrococcal nuclease (9).

For DNase I footprinting, we recommend using a biotinylated primer 1 and enriching for the primer extension products by capture on streptavidin-coated paramagnetic beads (35). This method, which is described in *Protocol 6*, greatly improves specificity, and for some primers this can be the key to obtaining LMPCR results of adequate quality.

It is essential to compare the *in vivo* DNase I ladder pattern with DNA from the same cells cleaved *in vitro*. This is because the pattern of 5-methylcytosine in the DNA will influence the pattern of DNase I cleavage (9, 29).

Protocol 2. *In vivo* footprinting procedures

Equipment and reagents

- PBS (*Protocol 1*)
- Solution 1: 150 mM sucrose, 80 mM KCl, 35 mM Hepes, pH 7.4, 5 mM K_2HPO_4, 5 mM $MgCl_2$, 0.5 mM $CaCl_2$
- Solution 2: 150 mM sucrose, 80 mM KCl, 35 mM Hepes, pH 7.4, 5 mM K_2HPO_4, 5 mM $MgCl_2$, 2 mM $CaCl_2$
- Solution 3: 150 mM NaCl, 5 mM EDTA, pH 7.8
- Lysolecithin, type I (Sigma).
- DNase I, pancreatic DNase I, Grade I (Worthington or Boehringer Mannheim)
- TE buffer: 10 mM Tris–HCl, pH 8.0, 1 mM EDTA. 1/10 TE: Ten-fold dilution of TE
- 100 mg/ml RNase A (DNase-free) in 10 mM Tris–HCl, pH 7.5 (Boehringer Mannheim; or prepared according to ref. 27)
- 99% Dimethyl sulfate (DMS) (Aldrich)
- Proteinase K (Boehringer Mannheim)
- DNase stop solution: 20 mM Tris–HCl, pH 8.0, 20 mM NaCl, 20 mM EDTA, 1% SDS, 600 μg/ml proteinase K (weigh and add just prior to use)
- DNase solution: 25 to 50 μg/ml DNase I in solution 2. (Prepare just before use. The concentration of DNase to use will vary with the cell line and other conditions and must be determined empirically.)
- 40 mM TNM buffer: 40 mM Tris–HCl, pH 7.7, 10 mM NaCl, 6 mM $MgCl_2$
- Reagents and apparatus for neutral and alkaline agarose gel electrophoresis[a]
- 10–15 cm Petri dishes

A. *In vivo footprinting using dimethyl sulfate*

1. Grow cells, preferably as a monolayer in a 10–15 cm Petri dish.
2. Prepare, at room temperature, a 0.2% solution of DMS in the cell growth medium without serum and immediately go to step 3.
3. Remove medium from the cells and replace it with 10 ml of freshly prepared 0.2% DMS solution.
4. Incubate at room temperature for 5–10 min[b].
5. Remove the DMS-containing medium and wash the cells with 10–20 ml of PBS.
6. Remove the cells by adding 5 ml of PBS and scraping the dish. Dilute with ice-cold PBS to 30 ml and centrifuge.

7. Wash the cells with 20 ml cold PBS.
8. Isolate nuclei and DNA as described in *Protocol 1*, minimizing incubation times to avoid depurination of 3-methyladenines.
9. Using *Protocol 3B*, starting at step 6, cleave 50–100 μg of this DMS-treated DNA with piperidine.
10. Determine the size of the fragments by alkaline 1.5% agarose gel electrophoresis and staining with ethidium bromide after neutralization[a].
11. If the DNA is the correct size (usually 300–600 nt), proceed to the LMPCR steps (*Protocol 4*).

B. *In vivo footprinting using DNase I*

1. Grow cells as monolayers to about 80–90% confluency in a 10 cm Petri dish or T75 flask. The procedure that follows is for $\sim 4 \times 10^6$ cells.
2. Remove the medium and cover the cells with 2.5 ml of 0.05% (w/v) lysolecithin[b] in solution 1.
3. Incubate for 1 min at 37 °C.
4. Remove the lysolecithin solution and rinse the cells with 10 ml of solution 1.
5. Add 2.5 ml of DNase solution and incubate at room temperature for 5 min[b].
6. Stop the reaction by removing the DNase I solution; then add 2.5 ml of DNase stop solution and 2.5 ml of solution 3.
7. Incubate for 3 h at 37 °C.
8. Complete the purification of the DNA by phenol–chloroform extraction and ethanol precipitation according to *Protocol 1*, steps 8–13. Dissolve DNA in TE or 1/10 TE buffer using 200 μl per 10^7 cells.
9. Remove RNA[c] in the DNA sample by treating with 50 μg/ml RNase in TE buffer for 1 h at 37 °C and repeating the phenol–chloroform extraction and ethanol precipitation. Dissolve the DNA pellet in water or 1/10 TE buffer and adjust the concentration to 1–2 μg/μl.
10. Determine the size of the fragments by alkaline 1.5% agarose gel electrophoresis and staining with ethidium bromide after neutralization[a].
11. If the DNA is the correct size (usually 300–600 nt), proceed to LMPCR. We recommend using the magnetic bead procedure described in *Protocol 6*. Alternatively, block the 3′ ends (9, 36) or try using the Vent procedure in *Protocol 5*.

C. *Naked DNA control*

1. Treat 40 μg of DNA[d] with 0.8 to 1.6 μg/ml DNase I for 10 min at room temperature in 40 mM TNM buffer.

Protocol 2. *Continued*

2. Stop the reaction by adding DNase stop solution. Extract with phenol–chloroform and ethanol-precipitate. Dissolve the DNA pellet in water or 1/10 TE buffer and adjust the concentration to 1–2 μg/μl.

[a] Solutions and procedures very commonly performed in molecular biology are as in Sambrook *et al.* (27) unless specified otherwise.
[b] The concentration of detergent and enzyme must be experimentally determined, as must the time of treatment.
[c] Samples for LMPCR should be substantially free of RNA
[d] DNA prepared as in *Protocol 1* is generally used.

4. Maxam–Gilbert chemical cleavage

A DNA sequence ladder is usually essential for the precise localization of footprints or adduct positions, so in most LMPCR experiments we include control lanes with genomic DNA cleaved at specific bases by the Maxam–Gilbert procedures (37). Maxam–Gilbert-cleaved DNA reliably gives clear LMPCR ladders with good signal strength, so it is an excellent DNA substrate to use for learning and testing the technique. Maxam–Gilbert-treated DNA tends to give a clean background. This is thought to be because the piperidine step, in addition to leaving a 5′ phosphate, also leaves a 3′ phosphate, blocking the 3′ ends and preventing them from functioning in the initiation of unwanted primer extensions.

The procedures described below in *Protocol 3* work well for 20–80 μg of DNA per base-specific reaction. If less DNA is available, reaction times can be shortened and/or temperature can be lowered. It is also possible to include carrier DNA (e.g. bacterial DNA) with the genomic DNA. When plasmid DNA is used, bacterial DNA (50 μg) should be added to dilute the target sequence.

The A-specific reaction described in *Protocol 3A*, step 5 works well for LMPCR, providing only A bands instead of A+T bands. The A-specific cleavage method given below is based on a protocol published by Iverson and Dervan (38), and was optimized for LMPCR by K. Graham and A. Riggs (unpublished).

Protocol 3. Chemical cleavage procedures

Equipment and reagents

- Dimethyl sulfate (DMS) 99+% pure (Aldrich). **Caution**: hazardous chemical! (see footnote *a*). Store under nitrogen at 4°C.
- DMS buffer: 50 mM sodium cacodylate, pH 8.0, 1 mM EDTA
- DMS stop solution: 1.5 M sodium acetate, pH 7, 1 M 2-mercaptoethanol
- 96% formic acid (Fluka)
- 1 M piperidine (Sigma or Fluka). Prepare fresh from 99% pure solution. Store undiluted stock under nitrogen at −20°C.

- Hydrazine: 98%, anhydrous (Aldrich). **Caution**: hazardous chemical! (see footnote *a*). Store under nitrogen at 4°C in an explosion-proof refrigerator and replace every 6 months.
- Hydrazine stop solution: 0.3 M sodium acetate, pH 7.5, 0.1 mM EDTA
- Lid locks for a microcentrifuge (National Scientific Supply)
- High-quality water
- Potassium tetrachloropalladate (K_2PdCl_4), 98% pure (Aldrich)
- 100 mM HCl, pH 2.0. Add 1.65 ml of reagent-grade HCl to 160 ml of high-quality water. Adjust the pH to 2.0 using 1 M NaOH; do not overshoot the pH. Add water to a total volume of 200 ml. This solution may be stored for several years.
- Speed-Vac rotary evaporator (Savant)
- Teflon™ tape

A. *Base-specific reactions*

1. Carry out the G-specific reaction:
   (a) Mix, on ice, 5 μl of genomic DNA solution (containing 50–100 μg of DNA in water), 200 μl of DMS buffer, and 1 μl of DMS[a].
   (b) Incubate at 20°C for 3–5 min.
   (c) Terminate the reaction by adding 50 μl of DMS stop solution. Mix by pipetting for a few seconds and then add 750 μl of cold (−70°C) ethanol.

2. Carry out the G+A-specific reaction:
   (a) Mix, on ice, 11 μl of genomic DNA solution (50–100 μg) and 25 μl of formic acid.
   (b) Incubate at 20°C for 10 min
   (c) Add 200 μl DMS stop solution. Mix by pipetting for a few seconds and then add 750 μl of cold (−70°C) ethanol.

3. Carry out the T+C-specific reaction:
   (a) Mix, on ice, 20 μl of genomic DNA solution (50–80 μg) and 30 μl of hydrazine[a].
   (b) Incubate at 20°C for 20 min.
   (c) Terminate the reaction by adding 200 μl of the hydrazine stop solution. Mix by pipetting for a few seconds and then add 750 μl of cold ethanol (−70°C).

4. Carry out the C-specific reaction:
   (a) Mix, on ice, 5 μl of genomic DNA (50–80 μg) and 15 μl of 5 M NaCl. (Mix well before adding the hydrazine!)
   (b) Add 30 μl of hydrazine[a].
   (c) Incubate at 20°C for 20 min.
   (d) Terminate the reaction by adding 200 μl of the hydrazine stop solution. Mix by pipetting for a few seconds and then add 750 μl of cold (−70°C) ethanol.

5. Carry out the A-specific reaction:
   (a) Make a fresh solution daily containing 10 mM K_2PdCl_4 in 100 mM HCl (pH 2.0). This solution is made by dissolving 3.26 mg of K_2PdCl_4 in 1 ml of 100 mM HCl.

Protocol 3. *Continued*

(b) Prepare a solution of 20 μg of genomic DNA in 160 μl of high-quality water.

(c) In a good fume hood initiate the reaction by adding 40 μl of the 10 mM K_2PdCl_4 solution prepared in step 5(a).

(d) Allow the reaction to proceed for 10 min at room temperature.

(e) Terminate the reaction by adding 50 μl of DMS stop solution. Mix by pipetting for a few seconds and then add 750 μl of cold (– 70°C) ethanol. Proceed to *Protocol 3B*, step 1 below. The pellet will be yellow, but is treated the same as the other DNAs.

B. *Purification and cleavage of treated DNA*

1. Keep all samples in a dry ice/ethanol bath (– 70°C) for at least 15 min.
2. Centrifuge for 15 min at ~ 14 000 g in a microcentrifuge at 0–4 °C.
3. Remove the supernatant, recentrifuge briefly, and carefully remove all liquid.
4. Resuspend the pellet in 225 μl of water. Make sure the pellet is dissolved!
5. Ethanol-precipitate using the following procedure. Add one-tenth volume (225 μl) of 3 M Na-acetate, pH 5.2, and then 2.5 volumes (750 μl) of ethanol pre-cooled to – 70°C. Keep on dry ice for 15 min and then spin for 10 min at ~ 15 000 *g* in a microcentrifuge at 0–4°C. Remove the supernatant and respin to ensure that the pellet is firmly stuck to the tube wall. Wash the pellet with 1 ml 75% (v/v) ethanol and spin for 5 min at 15 000 *g* in a microcentrifuge at 0–4°C. Dry the pellet under vacuum in a Speed-Vac rotary evaporator (Savant).
6. Dissolve the pellet in 100 μl of 1 M piperidine.
7. Place Teflon™ tape under the lids and close them, so that the Teflon™ will act as a gasket to prevent piperidine from escaping. Lock the lids with Lid Locks (National Scientific).
8. Heat at 90°C for 30 min in a heat block
9. Transfer to new tubes if the caps do not seal properly due to stretching during the previous step.
10. Ethanol-precipitate as in step 5, spinning for 15 min at 15 000 *g* in a microcentrifuge at 0–4°C, and washing the pellet twice with 80% ethanol.
11. Remove any remaining traces of piperidine by drying the sample overnight in a Speed-Vac rotary evaporator.
12. Dissolve DNA in water and adjust the concentration to 0.5–1 μg/ml.

[a] Dimethyl sulfate (DMS) and hydrazine are highly toxic chemicals and potential carcinogens. Handle these chemicals and piperidine in a chemical fume hood. Detoxify hydrazine waste (including plastic material) in 3 M ferric chloride. Inactivate DMS waste in 5 M NaOH.

5. LMPCR using Sequenase™/*Taq*

5.1 Primer selection

One of the most important factors determining success with LMPCR is the gene-specific primer set. In all, three gene-specific primers must be used. Primer 1 is used for the primer extension step, primer 2 for the PCR step, while primer 3 is used to make a hybridization probe. The three (nested) primers must be in close proximity or, preferably, slightly overlapping. If one of the primers is poor, then the entire procedure will also be poor. One should try to avoid repetitive sequences—and repetitive DNA sequences are common, making up over 10% of the mammalian genome (39, 40). A database containing a collection of consensus sequences for repetitive DNAs has been compiled by A. F. Smit and J. Jurka, and it is available by anonymous FTP from ncbi.nlm.nih.gov in the subdirectory repository/repbase/REF. There are some additional constraints on primer selection. For the Sequenase™/*Taq* procedure in *Protocol 4*, the first primer must be specific at 48 °C, because a key feature is primer extension at 48°C. This is on the high side of the temperature range for Sequenase™ 2.0, but is important because it limits the addition of an extra A overhang (unpublished data), which prevents ligation of the linker. Primer 2 must work together with the linker-primer in the PCR step. Thus for the standard protocol and linker, one needs to select a primer 2 with a T_m of 63–70 °C. As is general for PCR, this primer must not self-prime or have homology with the other primer, in this case the linker-primer.

We have had good success using the OLIGO™ program (National Biosciences, ref. 41) to aid the design of primers. We use program settings of 50 mM salt and 250 pM oligonucleotide concentrations, the default settings of the program and the settings used for published PCR studies (41). We have found that about 80–90% of primer sets designed with the aid of OLIGO™ work adequately well. Since it takes several days to obtain LMPCR results and the procedure has many potential variables, one may want to make an additional PCR primer(s) upstream and on the opposite strand, so that conventional PCR can be used to check each individual primer—two primer sets on opposite strands, as illustrated in *Figure 1B*, are convenient for this purpose.

Primer choice becomes even more critical for multiplexing, the simultaneous use of more than one gene-specific primer (42). More than one primer set can be included in the primer extension and PCR steps, thus simultaneously amplifying fragments from different genes or adequately separated regions of the same gene. For well-characterized genes, we have, in fact, routinely revealed sequentially two different sequences by stripping and rehybridization of the membrane (8). However, in becoming familiar with the technique and a new gene, one must be cautious about multiplexing because 'primer interference' is sometimes seen; that is, the inclusion of a second primer set can cause poor results for both.

5.2 Primer extension with Sequenase™

The oligonucleotides we have used as primer 1 have been 15- to 22-mers with a calculated T_m of 48–56°C. T_m is calculated using a computer program (OLIGO™, ref. 41) as discussed above. For many experiments involving comparisons between lanes, it is desirable to have roughly the same band strengths in each lane. To achieve this, the amount of DNA added to each reaction should be accurately known, either from absorbance, fluorescence measurement, or gel electrophoresis and comparison to standards. The latter, gel method is the most informative, because only DNA fragments within the range of primer extension contribute to the bands seen. The amount of DNA in the LMPCR reaction can be varied, but this should be done with awareness of potential problems at both the high and low end. For many DNAs, at least 5 μg of DNA per reaction can be used to obtain good signal strength, but, for reasons not fully understood, some DNAs show anomalous suppression of only certain bands at or above 5 μg per reaction (24). We have not seen this artefact when only 1 μg per reaction is used. For different reasons, one should be cautious about using less than 1 μg per reaction. This is because statistical sampling problems begin to appear when one uses less than 1 μg and the primers are of poor efficiency. Occasionally, bands will be missing or greatly reduced. However, footprint information can be obtained with as little as 0.1 μg/reaction, if multiple, independent samples are run and one confirms that the overall footprint pattern is reproducible (see section 10, Discussion).

5.3 Ligation

The primer-extended molecules which have a 5′ phosphate group are ligated to an unphosphorylated synthetic double-stranded linker. Because the linker is unphosporylated, it is essential for the 5′ strand of the genomic DNA to be phosphorylated. Maxam–Gilbert chemistry and DNase I procedures (*Protocols 3* and *2*, respectively) provide phosphorylated ends. Other procedures do not; micrococcal nuclease, for example, produces 5′ hydroxyls and 3′ phosphates. If the 5’ end is not phosphorylated, then a kination step must be added to the procedure (9).

The linker we usually use is the one originally described by Mueller and Wold (6). It works well as a linker and the longer strand subsequently works well as a primer, even though it is G+C rich and has internal symmetry. We have tested various other linkers and most work, but none have been better than the original one. Importantly, the longer oligonucleotide (25-mer) of the linker, which is used as a primer for PCR, should be gel-purified and/or show only a single band, as LMPCR provides single base-pair resolution.

5.4 PCR amplification

As with any PCR procedure, care has to be taken to avoid carry-over of PCR products. We perform all steps *before* PCR using pipette tips with aerosol barriers (USA Scientific Plastics), and specially designated pipettes and solutions. Pipette shafts are soaked in 1 M HCl if contamination is suspected.

The gene-specific primer used in the amplification step (primer 2) has usually been 20–28 nucleotides long, with a T_m calculated by the OLIGO™ program to be between 63 and 70 °C (41). It is designed to be downstream (nested) with respect to primer 1. As performed originally (6), we try to design primer 2 to overlap several bases with primer 1. This is not likely to be essential, but the idea is to provide competition with any residual primer 1 that may be carried with the sample to the PCR step. We have had good results with a second primer that overlaps only two bases with the first. The linker-primer is the longer oligonucleotide (the 25-mer) of the linker; its sequence is given in *Protocol 4*. PCR amplification primers, both linker-primer and gene-specific, should be gel-purified (43) and/or show only a single band.

We routinely perform the annealing step of PCR at the T_m of primer 2. According to the OLIGO™ primer analysis program, the linker-primer has a T_m of 67 °C. The temperatures given in *Protocol 4* may need to be adjusted for some primers. Importantly, only 18–20 cycles are usually performed, to ensure enzyme excess and to minimize preferential amplification of short sequences.

To completely extend all DNA fragments and uniformly add an extra nucleotide by the terminal transferase activity of *Taq* polymerase, an additional 'booster' step is part of the standard procedure. This step can be important and should not be omitted, otherwise double bands may be seen because only some molecules have an extra 3′ base.

Protocol 4. Ligation-mediated PCR using Sequenase™/*Taq*

Equipment and reagents

- 0.2 pmoles/μl primer 1 stock solution. Prepare in water, assuming that 1 A_{260} unit = 30 μg and 1 pmol = 330 pg × length of oligonucleotide. Store frozen.
- 20 pmoles/μl linker stock solution. The linker consists of a 25-mer (5′-GCGGTGACCCGGGAGATCTGAATTC) annealed to an 11-mer (5′-GAATTCAGATC) as follows. Prepare a solution containing the 25-mer and 11-mer (each at 20 pmoles/μl) in 250 mM Tris–HCl, pH 7.7. Heat to 95°C for 3 min, then at 70 °C for 1 min, and then gradually cool to 4°C over a period of 2 h. The linker stock solution can be stored at −20°C for at least 3 months. Keep the solution on ice after thawing.
- 50 pmoles/μl primer 2 stock solution. Prepare in water. Calculate concentration as for primer 1.
- Bovine serum albumin (BSA) (New England Biolabs)
- 50 pmoles/μl linker-primer stock solution. The linker-primer is the longer strand of the linker. Store the stock solution frozen.
- 13 units/μl Sequenase™ 2.0 (United States Biochemical). We do not use Sequenase 1.0.
- 3 units/μl T4 DNA ligase (Promega)
- 10 mg/ml *E. coli* tRNA in water (Boehringer Mannheim). Store frozen.
- TE buffer, pH 8.0 (*Protocol 1*)
- 5 × Sequenase™ buffer: 250 mM NaCl, 200 mM Tris–HCl, pH 7.7 at room temperature
- 100 mM ATP
- 25 mM dNTP. Prepare by mixing equal volumes of 100 mM stock solutions of dATP, dGTP, dCTP, and dTTP.
- Ligation stop mix. For 20 samples mix 168 μl of 3 M Na-acetate, pH 5.2, and 20 μl of *E. coli* tRNA (10 mg/ml).
- Phenol/chloroform

Protocol 4. *Continued*

- 5 × *Taq* buffer: 200 mM NaCl, 50 mM Tris–HCl, pH 8.9, 0.05 % (w/v) gelatin
- 5 units/μl *Taq* polymerase (AmpliTaq™; Perkin Elmer, or Boehringer Mannheim)
- 1 M DTT
- *Taq* stop mixture. For 20 samples, mix 250 μl 3 M Na-acetate, pH 5.2, 50 μl 0.5 M EDTA, pH 7.7, and 10 μl 10 mg/ml *E, coli* tRNA.
- Siliconized 0.65 ml microcentrifuge tubes
- Speed-Vac rotary evaporator (Savant)

A. *Primer extension*

1. Mix in a siliconized 0.65 ml microcentrifuge tube:
   - H_2O to make final volume 15 μl
   - 3 μl 5 × Sequenase™ buffer
   - 3 μl primer 1 (0.6 pmoles)
   - 0.5–2 μg DNA in TE buffer or water
2. Denature at 95°C for 3 min.
3. Anneal at 45°C for 30 min.
4. Chill on ice and spin briefly.
5. Prepare Mg–dNTP solution:

Stock solution	*Vol.*	*Final conc.*
• water	380 μl	
• 1 M $MgCl_2$	8 μl	20 mM
• 1 M DTT	8 μl	20 mM
• 25 mM dNTP	4 μl	0.25 mM each dNTP

6. Add 7.5 μl of cold, freshly prepared Mg–dNTP solution to each reaction.
7. Dilute stock Sequenase™ (13 units/μl) fourfold in ice-cold TE buffer, pH 8.0, to give 3.25 units/μl.
8. Add 1.5 μl (~ 5 units) of the freshly diluted Sequenase™ solution to the reaction mixture.
9. Incubate at 48°C for 15 min.
10. Cool on ice.
11. Add 6 μl of 300 mM Tris–HCl, pH 7.7.
12. Heat-inactivate at 67°C for 15 min.
13. Cool on ice and spin briefly.

B. *Ligation*

1. Prepare a ligation mixture containing:

Stock solution	*Vol,/tube*	*20 tubes*	*Final conc.*
• water	35.55 μl	11 μl	
• 1 M $MgCl_2$	0.60 μl	12 μl	13.33 mM
• 1 M DTT	1.35 μl	27 μl	30 mM
• 100 mM ATP	0.75 μl	15 μl	1.66 mM
• 5 mg/ml BSA	0.75 μl	15 μl	83.3 μg/ml
• Linker (20 pmol/μl)	5.0 μl	100 μl	100 pmoles/reaction
• T4 DNA ligase	1.0 μl	20 μl	3.0 units/reaction

2. Add 45 μl of the freshly prepared ligation mix to the DNA samples from part A, step 13 above.
3. Mix and incubate overnight at 18–20 °C.
4. For DNase I treated samples, go to *Protocol 6*, otherwise continue.
5. Incubate for 10 min at 70 °C to inactivate the ligase. Spin briefly.
6. Add 9.4 μl of ligation stop mix.
7. Ethanol-precipitate and wash as follows. Add 220 μl of 100% ethanol, chill on dry ice for 20 min, and spin at 4 °C for 15 min in a microcentrifuge at 15000 *g*. Remove the supernatant and wash the pellet with 500 μl 75% ethanol. Spin for 5 min at 4 °C in a microcentrifuge at 15000 *g*. Carefully remove the supernatant. (The pellet is transparent and hard to see.) Respin.
8. Remove residual ethanol in a Speed-Vac rotary evaporator.
9. Dissolve the pellet in 50 μl of H_2O.

C. *PCR amplification*

1. Prepare a 2 × *Taq* polymerase mixture:

	Per tube	*20 tubes after step 2*	*Final conc,*
• Water	28 μl	560 μl	
• 5 × *Taq* buffer	20 μl	400 μl	1 × *Taq* buffer
• 1 M $MgCl_2$	0.2 μl	4.0 μl	2 mM
• 25 mM dNTP mix	0.8 μl	16 μl	0.2 mM each dNTP
• primer 2 (50 pmol/μl)	0.2 μl	4.0 μl	10 pmoles/100 μl
• linker-primer (50 pmol/μl)	0.2 μl	4.0 μl	10 pmoles/100 μl
• *Taq* polymerase	0.6 μl	12 μl	3 units/100 μl

2. Add 50 μl of freshly prepared *Taq* polymerase mixture to the samples from part B, step 9 above, and mix by pipetting.
3. Cover the samples with 50 μl of mineral oil and spin briefly to separate the phases.
4. Thermal cycle 18–20 times with a program of 1 min at 95 °C, 2 min at the calculated T_m of primer 2 (60–66 °C) for 2 min, and 3 min at 76 °C.
5. Prepare a fresh booster mixture containing 1 unit of *Taq* polymerase per 10 μl of 1 × *Taq* polymerase mixture.
6. Add 10 μl of the booster mixture to each sample[a], mix and spin quickly.
7. Incubate at 74 °C for 10 min.
8. Chill on ice.
9. Add 15.5 μl of *Taq* stop solution to each sample, mix, and spin briefly.

Protocol 4. *Continued*

10. Transfer the sample from under the oil to a new tube[b].
11. Extract with 70 μl of phenol and 120 μl of chloroform (pre-mixed) to remove proteins and any residual oil.
12. Transfer the supernatants to new tubes.
13. Ethanol-precipitate, and wash the pellet as in part B, step 7 of this procedure.
14. Remove residual ethanol in a Speed-Vac rotary evaporator.

[a] Do not skip this step. The booster step is important to ensure that all molecules will have one extra nucleotide as a 3′ overhang. LMPCR fragments produced by the Sequenase/*Taq* procedure are uniformly one nucleotide longer than those produced by the $Vent_R$ (exo^-)/$Vent_R$ procedure.

[b] Alternatively, one can add the phenol/chloroform directly without removing the oil. This causes the aqueous layer to be the top layer.

6. Ligation-mediated PCR using $Vent_R$ polymerase

If the Sequenase™/*Taq* procedure is found to give weak or missing bands and/or a ladder that prematurely terminates, the use of alternative enzymes for LMPCR may be advantageous. Garrity and Wold (44) and others (33, 45) have reported on the use of recombinant Vent polymerase ($Vent_R$) for LMPCR, and we have also obtained good results, even though this enzyme has an exonuclease activity that degrades the primer during the polymerization reaction (confirmed by our unpublished results). This adds another variable which may depend on the efficiency of the primer. *Protocol 5* describes a procedure adapted from Garrity and Wold (44), Quivy and Becker (45), and Hornstra and Yang (33). The procedure describes the use of $Vent_R$ (exo^-), an exonuclease minus version of $Vent_R$, for extension of primer 1 and then regular $Vent_R$ for the PCR step. In our hands the procedure described by Garrity and Wold, which uses $Vent_R$ for both the first primer extension and PCR, often works well but is not as reliably robust as is the Sequenase™/*Taq* procedure of *Protocol 4*, or the $Vent_R$ (exo^-)/$Vent_R$ procedure of *Protocol 5*. For unknown reasons, Sequenase™/*Taq* seems to give stronger signals than do other procedures for DNA having a low density of nicks (e.g. one nick per 6 kb), even though band strength and clarity is about the same using Maxam–Gilbert cleaved genomic DNA. As for Sequenase™ and *Taq*, $Vent_R$ (exo^-) is known to add an extra A to a percentage of 3′ ends, creating an unligatable overhang. Nevertheless, our experimental observation is that, as first described by Quivy and Becker (45), $Vent_R$ (exo^-) works well for LMPCR. We are presently using $Vent_R$ (exo^-) mainly for problematical sequences that terminate or give missing bands with Sequenase™. Problems seem prone to occur in homopurine stretches, although a clear understanding and the ability to predict has not yet emerged.

Protocol 5. Ligation-mediated PCR using $Vent_R(exo^-)/Vent_R$

Equipment and reagents

- 2 units/μl Vent polymerase, recombinant exonuclease minus ($Vent_R$ (exo^-)) (New England Biolabs)
- 2 units/μl Vent polymerase, recombinant ($Vent_R$) (New England Biolabs)
- 10 × Vent buffer, pH 8.8: 100 mM KCl, 100 mM $(NH_4)_2SO_4$, 200 mM Tris–HCl, pH 8.8 at 25°C, 20 mM $MgSO_4$, 1% Triton X-100 (supplied by New England Biolabs)
- 100 mM $MgSO_4$
- Freshly prepared vent dilution solution: 44 μl 1 M Tris–HCl, pH 7.5, 7.2 μl 1 M $MgCl_2$, 20 μl 1 M DTT, 5 μl BSA (10 mg/ml), in a final volume of 400 μl
- Freshly prepared ligation buffer (*Protocol 4B*). Adjust the $MgCl_2$ concentration so that the final magnesium concentration in the ligation reaction is close to 10 mM.
- Siliconized 0.65 ml microcentrifuge tubes
- Lid locks (*Protocol 3*)

Method

1. Add to a siliconized 0.65 ml microcentrifuge tube on ice:
   - 1 μl of DNA (1 μg)
   - 3 μl of 10 × $Vent_R$ buffer
   - 1.2 μl of 100 mM $MgSO_4$
   - 3 μl of primer 1 (0.6 pmoles)
   - 1.5 μl of 5 mM dNTPs
   - 19.3 μl H_2O
2. Mix by pipetting, then give a quick spin.
3. Denature in a thermal cycler at 96°C for 8 min with lid locks on the tubes. Place on ice and give a quick spin at 4°C.
4. Anneal at 54–60°C for 20–30 min. The annealing temperature should be 0–5°C above the T_m of primer 1, as determined by the use of the OLIGO™ 4.0 program for standard PCR conditions[a].
5. Fast spin in a microcentrifuge at 4°C, then add, on ice, 1 μl of $Vent_R$ (exo^-) polymerase (2 units).
6. Carry out the primer extension first for 3 min at 0–5°C above[a] the T_m of primer 1, and then for 3 min at 72°C. Alternatively, a ramp-up procedure in a thermocycler can be used[a].
7. Fast spin and add 20 μl of ice-cold vent dilution buffer and 25 μl of ice-cold ligation buffer. (These two solutions may be mixed and added together.)
8. Mix and incubate for 4 h or overnight at 17–18°C. Put samples on ice. Quick spin at 4°C. If biotinylated primer 1 is used, samples are processed as described in *Protocol 6*.
9. Add 40 μl of 7.5 M ammonium acetate and 1 μl of 20 mg/ml glycogen to each tube, followed by 2 volumes of ethanol.

Protocol 5. *Continued*

10. Incubate on dry ice for 20 min, centrifuge for 30 min at 4 °C in a microcentrifuge at 14 000 *g*, wash with 400 μl 80% ethanol, centrifuge for 6 min, and air-dry the pellet.
11. Dissolve the pellet in 50 μl water. Keep on ice.
12. To each tube add:
    - 10 μl of 10 × $Vent_R$ buffer
    - an amount of 100 mM $MgSO_4$ determined empirically[b] (the final concentration is usually 2–6 mM $MgSO_4$)
    - 1 μl 25 mM dNTPs
    - 0.4 μl of 50 μM linker-primer
    - 0.4 μl of 50 μM primer 2
    - 1 μl of $Vent_R$ (1 U/μl)
    - H_2O to give a final volume of 100 μl.
13. Mix, quick spin, and add 50 μl of mineral oil to each tube.
14. Add samples to a thermal cycler at 80 °C. Typical cycling conditions for a primer 2 with a calculated T_m of 67 °C are: one cycle of 95 °C for 3.5 min, 66 °C for 2 min, 76 °C for 3 min; 18 cycles of 95 °C for 1 min, 66 °C for 2 min, 76 °C for 3 min; one cycle of 95 °C for 1 min, 66 °C for 2 min, 76 °C for 10 min.
15. Process the amplified samples as in *Protocol 4C*, starting at step 8.

[a] The OLIGO™ program seems to underestimate somewhat the effective T_m under $Vent_R$ buffer conditions. For primers originally designed for the Sequenase/*Taq* procedure the calculated T_m will be about 55 °C, a temperature at which $Vent_R$ (exo^-) is not very active. To compensate for this and to ensure complete elongation, we incubate first at the lower temperature and then at a temperature closer to the optimum for the enzyme. A number of successful experiments have been performed using a thermal cycler and a ramp-up program of 55 °C for 1 min, 57 °C for 1 min, 60 °C for 1 min, 64 °C 1 min, 68 °C 2 min, 72 °C for 2 min, and 76 °C for 3 min.

[b] The optimum magnesium concentration for $Vent_R$ polymerase is higher than for *Taq* polymerase, and it can be as high as 8 mM. As for the Sequenase/*Taq* procedure, it is advisable that preliminary PCR experiments be performed to check the primer and to determine the best conditions.

7. Extension-product capture and enrichment

DNase I treated DNA often gives poor quality ladders relative to the same DNA cleaved by Maxam–Gilbert chemistry. The cause for the lower specific signal and higher background is thought to be that DNase I cleavage produces DNA strands with 3′ hydroxyl ends. These ends can act as primers for, in effect, random primer extension, and thus a large number of molecules can participate in linear PCR, leading to a higher background. Some primer sets are so robust that good absolute band strength and signal-to-background ratios can be obtained without special treatment. This may be especially true if the Vent procedure is used (44). However, the efficiency of primer sets is quite variable, so methods to improve the signal-to-noise ratio are often needed. Blocking the 3′ ends by adding a dideoxynucleotide using terminal transferase (9) gives some improvement, but we now use a procedure which enriches for the products of the primer extension reaction (35). The procedure employs a 5′ biotinylated primer 1, so that molecules incorporating primer 1 are captured by streptavidin-coated paramagnetic beads. Other genomic DNA fragments will remain in solution and be removed when the beads are washed. This magnetic bead procedure is given in *Protocol 6*. It should be noted that the 'old' template strand is the one subsequently used for PCR. Excess primer 1 and the newly synthesized strand containing the biotinylated primer remain attached to the beads even in alkali and are discarded.

Protocol 6. LMPCR using extension product capture and enrichment

Reagents

- Streptavidin-coated paramagnetic beads (Dynabeads M280, 6–7 × 10^8 beads/ml; Dynal)
- 2 × binding and washing (BW) buffer: 10 mM Tris–HCl, pH 7.5, 1 mM EDTA, 2 M NaCl
- 0.15 M NaOH. Prepare this solution fresh, to minimize a molarity change due to CO_2 absorption from the air.
- 2 M Tris–HCl, pH 7.7
- 0.15 M HCl
- Biotinylated primer 1. (This primer is the same as for standard LMPCR except that the oligonucleotide has biotin attached by a nine-atom linker arm to the 5′ terminus. Synthesis of our biotinylated oligonucleotides is achieved on an automated synthesizer using the phosphoramidite 1-dimethoxytrityloxy-2-(*N*-biotinyl-4-aminobutyl)-propyl-3-*O*-(2-cyanoethyl)-*N*,*N*-diisopropyl)-phosphorami dite obtained from Glen Research.)

Method

1. Using a biotinylated primer 1, perform the primer extension and ligation steps of *Protocol 4B*, through step 3.
2. Prepare paramagnetic beads (37.5 μl/sample) by washing twice with 2 × BW buffer (75 μl/sample each time) and resuspending the beads in 75 μl 2 × BW buffer/sample. Wash just before use.
3. Add the ligation mixture (75 μl) to the washed beads.
4. Incubate for 15 min at room temperature, keeping the beads suspended by gently rocking the tubes to prevent the beads from settling[a].
5. Using a magnet to hold the paramagnetic beads on the side of the tubes, remove the supernatant and wash the beads twice with 75 μl of 2 × BW buffer.
6. Add 37 μl of 0.15 M NaOH and incubate at 37 °C for 10 min. The non-biotinylated ('old') template DNA strand will go into solution.
7. Use a magnet to hold the beads on the wall of the tubes and transfer the alkaline solution to new tubes[b].

Protocol 6. *Continued*

8. Neutralize the alkaline solution by adding first 3.75 μl of 2 M Tris–HCl, pH 7.7, and then 37 μl 0.15 M HCl; use the same pipette and setting as for the NaOH solution when adding the HCl.
9. Proceed to *Protocol 4B*, step 6 and continue the LMPCR procedure.

[a] Enough beads must be used to adsorb the biotin-tagged primer as well as the extension-product molecules. Since the beads settle rather rapidly, it is probably important to keep them suspended.
[b] Single-stranded DNA in alkali can be adsorbed to and thus lost on the walls of polypropylene tubes, even siliconized tubes. Although we have no documentation that this is a problem for this procedure, it is nevertheless recommended that the time in alkali be kept to a minimum.

8. Gel electrophoresis, transfer to nylon membrane, and UV fixation

As first carried out by Mueller and Wold (6), many investigators visualize the LMPCR ladder by performing, after the PCR step, a primer extension reaction using primer 3 labelled with ^{32}P at the 5′ end by kination. This method does have the advantage that two steps, transfer to membrane and hybridization, can be avoided. However, we generally prefer electroblotting and hybridization. A longer hybridization probe is generally more specific than an oligomer, and the hybridization signal increases proportionally with the size of the membrane-bound LMPCR fragments. Also, nylon membranes can be rehybridized, and different gene-specific ladders can be revealed if multiple primer sets are used for the primer extension and PCR reactions (multiplexing). Most importantly, though, for laboratories that do many LMPCRs, the hybridization approach as described in *Protocol 8* results in significantly less radiation exposure.

Protocol 7. Gel electrophoresis, transfer to membrane, and UV fixation

Equipment and reagents

- 0.5 M TBE buffer, pH 8.3: dissolve 60.55 g Tris base, 25.7 g boric acid, and 1.85 g EDTA-dihydrate in 1 litre of distilled water. Store at room temperature.
- 8% (w/v) acrylamide, 7 M urea, 0.1 M TBE buffer: weigh out 38.65 g acrylamide (**Caution**: toxic, use a mask!), 1.35 g bisacrylamide (**Caution**: use a mask), and 210 g urea. Dissolve in 100 ml 0.5 M TBE buffer, pH 8.3, and add water to a final volume of 500 ml. Degassing is unnecessary.
- 10% ammonium persulfate: dissolve 1 g of ammonium persulfate in 9 ml of water. Store at 4°C and make fresh daily.
- Formamide loading solution: 95% (v/v) formamide, 20 mM EDTA, pH 8.0, 0.05% (w/v) xylene cyanol, 0.05% (w/v) Bromophenol blue
- DNA sequencing electrophoresis apparatus and power supply (4000 W or greater). Any system that is adequate for DNA sequencing is adequate for LMPCR.

- Temperature indicator strip for gel plates (Bio-Rad)
- Flat pipette tips: Multi Flex 0.4 mm (Sorenson Bioscience)
- Electrotransfer apparatus. We are presently using an electrotransfer apparatus obtained from Owl Scientific (Semi-dry electroblotter and DNA sequence transfer system, model HEP-3); however, previously we used an easily home-constructed apparatus. This apparatus, which can be cheaply made using stainless-steel plates (obtainable as spare parts from Hoefer Scientific) from a Bio-Rad gel dryer as the electrodes, is fully described in ref. 43. The main advantage of the Owl electroblot apparatus is that it needs much less transfer solution and 3 MM paper.
- UV energy monitor (UVX radiometer; UV Products)
- *N,N,N′,N′*-tetramethylethylenediamine (TEMED) (Sigma)
- UV cross-linking apparatus. UV Stratalinker™ (model 2400, Stratagene) or, alternatively, an inverted, suspended transilluminator with the upper lid removed. The transilluminator holds six 254 nm germicidal 15 W UV tubes, and is usually positioned 20 cm above the membrane.
- Urea, ultrapure (ICN Biomedicals)
- Acrylamide and N,N′-methylene-bisacrylamide (bisacrylamide) (Bio-Rad)
- Whatman 3MM and No.17 chromatography paper, cut to 47 × 19 cm.
- Nylon filters, cut to 43 × 18 cm. GeneScreen™ membranes (Dupont/New England Nuclear) give good results. Membranes from other companies have worked well. There may be some lot to lot variation, so this should be kept in mind and checked if results are poor.
- Hybridization oven (Red Roller II, model HB 1100D; Hoefer Scientific)
- Saran wrap

A. *Preparation, loading, and running an acrylamide gel*

1. Prepare an 8% acrylamide, 7 M urea, 0.1 M TBE solution.
2. Mix 60 ml of acrylamide solution with 600 μl of 10% ammonium persulfate and 15 μl of TEMED.
3. Prepare an acrylamide gel that is 0.4 mm thick, 17 cm wide, and 60 cm long, and allow it to polymerize for at least 2 h.
4. Remove the comb, wash the wells with gel running buffer (0.1 M TBE), and pre-run the gel at 75 W for at least 1 h and until the temperature reaches 50 °C, as measured by a temperature indicator strip stuck to the glass plate.
5. Prepare samples by dissolving DNA pellets in 2 μl of water and adding 4 μl of formamide loading solution.
6. Just before loading the gel, heat the sample at 95 °C for 3 min and then chill on ice.
7. Load 3 μl of sample onto the gel using a thin, flat pipette tip.
8. Carry out the electrophoresis at 75 W constant power (~ 50 °C gel temperature) until the Bromophenol blue has run off the end of the gel and the xylene cyanol marker has reached the bottom (about 4 h).

B. *Electroblotting to a nylon membrane and UV fixing*

1. After the electrophoresis run, transfer the lower part of the gel (including the xylene cyanol dye and about 40 cm above it) to Whatman 3MM paper and cover the gel with Saran wrap. The size of the gel piece is 43 × 17 cm.
2. Prepare the electroblot apparatus (Owl Scientific) by placing on it three layers of Whatman No.17 paper pre-wet with TBE gel running buffer.

Protocol 7. *Continued*

Remove any bubbles in the paper stack, if necessary by rolling with a bottle or other suitable object. Uniformly distribute 50 ml of the running buffer onto the paper pile.

3. Carefully place the 3MM paper and adhering gel piece, still covered with Saran wrap, on the paper pile. Remove any air bubbles between the gel and paper by wiping over the Saran wrap with a soft tissue. When all air bubbles have been squeezed out, remove the Saran wrap and cover the gel with a wet nylon membrane (e.g. GeneScreen™) cut somewhat larger than the gel piece and pre soaked in gel running buffer. Put three layers of pre-soaked Whatman No.17 paper on top of the membrane, again being careful to avoid bubbles.
4. Transfer for 45 min using 12 V with a current limit of 1.6 amps.
5. Mark the DNA side of the membrane with a pencil before removing.
6. Dry the membrane briefly at room temperature.
7. Fix the DNA to the membrane by UV irradiation. If the Stratalinker is used, the Auto Crosslink setting is used. If the inverted illuminator is used, a UV dose of 1000 J/m^2 is given (about 35 sec at 20 cm from six 15 W UV bulbs. Since output varies with the age of the bulbs, check the UV dose with a UV monitor. Only the DNA side of the membrane is irradiated. Calibrating the UV irradiation time for different batches of GeneScreen™ membranes was found to be unnecessary (more than 10 different batches have been tried).

9. Probe synthesis and hybridization

The most convenient method for preparing hybridization probes is by repeated primer extension using *Taq* polymerase and a single primer (primer 3) on a double-stranded template, which can be either cloned DNA or PCR-derived DNA. The length of these probes can be easily controlled by the choice of primers or by cutting the plasmid template with a restriction enzyme. To minimize background, the probe should not include the sequence of primer 2 or its complement. Primer 3 should have a T_m of 60–68 °C, be nested inside primer 2, and bind to the same strand as primer 2 (see *Figure 1*). To minimize unwanted hybridization, primer 3 should not overlap primer 2 by more than 8–10 bases.

Protocol 8. Probe synthesis and hybridization procedures

Equipment and reagents

- 20 pmol/μl primer 3 in water
- [^{32}P]dCTP (3000 Ci/mmol; New England Nuclear)
- 0.5 M phosphate buffer, pH 7.2
- 20% (w/v) SDS in water. Store at room temperature.

- Hybridization buffer: 0.25 M sodium phosphate, pH 7.2, 1 mM EDTA, 7% (w/v) SDS, 1% (w/v) BSA. To prepare 50 ml of this solution mix 25 ml 0.5 M phosphate buffer, pH 7.2, 17.5 ml 20% SDS, 0.5 g BSA, 0.1 ml 0.5 M EDTA, and 7.4 ml water. To completely solubilize all components, heat to 50°C with stirring.
- Washing buffer A: 20 mM sodium phosphate, pH 7.2, 1 mM EDTA, 2.5 % (w/v) SDS, 0.25 % (w/v) BSA. Prepare by adding 40 ml 0.5 M phosphate buffer, pH 7.2, 125 ml 20% SDS, 2 ml 0.5 M EDTA, and 2.5 g BSA to 833 ml distilled water.
- 5 × *Taq* buffer (*Protocol 4*)
- 100 mM $MgCl_2$
- Washing buffer B: 20 mM sodium phosphate, pH 7.2, 1 mM EDTA, 1% (w/v) SDS. Prepare by adding 40 ml 0.5 M phosphate buffer, pH 7.2, 50 ml 20% SDS, and 2 ml 0.5 M EDTA to 908 ml distilled water.
- 1 mM each dATP, dGTP, dTTP
- *Taq* polymerase
- G-50 spin columns (Select D; 5 Prime-3 Prime)
- Hybridization oven with 250 ml hybridization cylinders (Red Roller II, model HB100D; Hoefer Scientific)
- Clinical centrifuge fitted with swinging-bucket rotor
- Saran wrap
- Kodak XAR-5 X-ray film

A. *Probe synthesis*

1. Prepare the template. Select a fragment or plasmid with the target gene, and, if necessary, cut with an appropriate restriction enzyme. Check completeness of cutting by agarose gel electrophoresis, then phenol–chloroform extract and ethanol precipitate.
2. Prepare a probe synthesis reaction by mixing, in order, at room temperature:

   water to make 100 μl final volume

• 5 × *Taq* buffer (*Protocol 4*)	20 μl
• 100 mM $MgCl_2$	2 μl
• 1 mM dATP	1 μl
• 1 mM dGTP	1 μl
• 1 mM dTTP	1 μl
• template	10–20 ng of the restriction-cut plasmid DNA (or 5 ng of a restriction fragment or PCR product)
• primer 3	1 μl (20 pmoles)
• [^{32}P]dCTP	5 μl (100 μCi)
• *Taq* polymerase	1 μl (5 units)

3. Cover samples with 50 μl of mineral oil.
4. Carry out 30 thermal cycles using a program of 95°C for 1 min, 60–68°C for 2 min, and 75°C for 3 min.
5. Remove sample from below the oil and transfer to a new tube.
6. Add 8 μl of 0.5 M EDTA and 10 μl of 10 mg/ml tRNA[a].
7. Prepare two G-50 spin columns by centrifuging in a swinging-bucket rotor at 1100 *g* for 1 min. Discard the eluate.
8. Clean the sample by adding no more than 50 μl per spin column, and centrifuging at 1100 *g* for 1 min. Collect the eluate.
9. Use the eluate as the hybridization probe within 2 days[b].

Protocol 8. *Continued*

B. *Hybridization procedure*

1. Soak the nylon membrane prepared in *Protocol 7* in 0.1 M TBE buffer and insert it into a hybridization cylinder. This can be done easily by rolling the membrane first onto a 25 ml pipette and then unspooling it into the cylinder. The membrane should stick completely to the walls of the cylinders without any air pockets. The membrane may overlap several times.
2. Pre-hybridize the membrane with 15 ml of hybridization buffer per cylinder for 15 min to 2 h in a hybridization oven at 60–68 °C.
3. Remove the pre-hybridization solution and add 5–7 ml of hybridization buffer and 100 μl (about 50 μCi) of the probe prepared in part A, step 9[b]. Roll the cylinder in the oven overnight at 60–68 °C. Although the theoretically optimum hybridization temperature will depend on the G+C content of the region, good results are usually obtained using a hybridization temperature of 60 °C.
4. Wash the filters. Remove a single membrane, or a maximum of two membranes at the same time, from the hybridization cylinders and wash first for 5 min in 250 ml of pre-warmed wash buffer A and afterwards for a total time of 30 min with a total of 2 litres of pre-warmed wash buffer B, 300–400 ml at a time, changing the buffer about 4 to 5 times (about every 7 minutes). Perform the washing steps at room temperature with buffers pre-warmed to 60 °C. Higher wash temperatures can be used if found to be necessary for regions of high G+C content.
5. After washing, dry the membranes at room temperature[c], wrap in Saran wrap, and expose to Kodak XAR-5 X-ray film (intensifying screens are generally used). Good intensity ladders are sometimes seen with as little as 0.5 h exposure. Most often exposure with intensifying screens for 2–10 h is optimum.

[a] Alternatively, the probe can be cleaned by phenol/chloroform extraction and ethanol precipitation using 7.5 M ammonium acetate.

[b] Radioactive DNA in hybridization buffer can be reused if within 2– 3 days.

[c] Nylon membranes can be sequentially rehybridized to reveal several gene-specific ladders if several sets of primers are included in the primer extension and amplification reactions. Strip the probe from the previous hybridization from the nylon membranes by soaking in 0.2 M NaOH for 30 min at 45 °C. If membranes are to be reused, it is best not to let them dry completely as this makes it more difficult to remove the probe.

10. Discussion

10.1 Reproducibility, quantitation, and sensitivity

Footprints and other chromatin features are identified in DNA-sequence-like ladders as regions of strong or weak bands relative to other bands above and/or below the footprint region. One must also compare the band pattern with that of naked DNA in an adjacent lane. Thus, successful identification of footprints requires reproducibility and depends on the quantitative aspects of LMPCR. Numerous experiments over a period of several years, often with identical samples prepared in parallel, have firmly established that for any given experiment the band pattern obtained is very reproducible. The reason that quantitation is adequate is that, although the efficiency of any given step may vary, all fragments are equally affected for any given sample. Importantly, during the PCR amplification step, all participating molecules have the identical 5′ and 3′ target sites for the PCR primers. The efficiency of amplification is still somewhat dependent on the sequence of a fragment, but band intensity relative to other nearby bands is an intrinsic property of each fragment. Although PCR is an exponential process, the amount of PCR product is linearly proportional to the template concentration (i.e. to the number of template molecules), as long as all other components of the reaction are in vast excess, as is usually the case for at least the first 20 cycles (46, 47). Theoretically, since nearby bands in the same lane act as internal controls, band patterns should be rather robust, and this has been confirmed experimentally. Lane-to-lane comparisons for experiments using about 1 μg of genomic DNA per reaction can reliably detect the twofold differences between one X chromosome-containing male DNA and two X chromosome-containing female DNA (8, 9).

The sensitivity of LMPCR is such that only a few molecules of each fragment size are needed. Because of this one must be cautious of statistical factors, as previously noted (7). We estimate that a single molecule that *participates* in primer extension and PCR can give a detectable LMPCR band. We further estimate that LMPCR can be about 10% overall efficient in molecule usage. Therefore as few as 10 molecules of a particular size can produce a band in the final gel ladder. However, if only 10 molecules are present in the sample, and efficiency is 10%, there is a severe statistical sampling problem. Thus, in order for most bands to be reliably represented, one needs at least 100 molecules of each size, and occasionally even then there will be a missing or very weak band just because of statistics. If 300 bands are observed, even rare statistical fluctuations, say at the 1% level, will cause occasional weak or completely missing bands. For these reasons, we recommend that each reaction tube of a DMS footprint experiment contains DNA from about 10^5 cells, although 10^4 cells may be adequate for excellent primer sets. DNase I footprinting very often gives clear footprints over a larger region, and for this type of experiment an occasional statistically missing band would be of little concern. However, DNase I experiments often need extra steps such as enrichment on magnetic beads, somewhat lessening any increase in sensitivity. As established by Steigerwald *et al.* (23), the use of restriction enzymes to cleave at specific sites greatly increases the sensitivity of LMPCR, because, in effect, the signal is concentrated into a single band instead of being spread over hundreds of bands.

10.2 Present applications and future prospects

There are now two well-established, LMPCR-based methods for obtaining *in vivo* footprint information in minimally perturbed cells. Treatment of intact cells with DMS has been most commonly performed (reviewed in ref. 14), and even though many contacts are missed, at least a few LMPCR bands are altered by most bound proteins. DNase I studies require a little more preliminary work to establish conditions for permeabilization, but very clear footprints can usually be obtained once these conditions are found. A third method, UV photofootprinting in conjunction with LMPCR (11), has recently been shown to have the potential to reveal all protein–DNA interactions, provided the factor binding site contains a dipyrimidine sequence (12). UV light has the advantage that perturbation is minimal and sensitivity can be excellent. Aside from the requirement of dipyrimidines in the site, this method is limited only by the need for special enzymes, T4 endonuclease V and *E. coli* photolyase, the latter of which is not yet commercially available.

LMPCR analysis of chromatin is still a young method and many additional improvements can be expected. To reveal regions of unusual *in vivo* DNA structure, for instance kinks, cruciforms, Z DNA/B DNA junctions, and single-stranded regions, various nucleases and agents that have been much used *in vitro*, such as $KMnO_4$, OsO_4, chloroacetaldehyde, and mung bean nuclease are compatible with LMPCR. $KMnO_4$, in particular, is convenient, gives excellent LMPCR ladders (9), and has been used for *in vivo* studies to identify regions of increased reactivity near transcription start sites (32, 48). LMPCR with micrococcal nuclease (MNase) has been used to reveal clearly positioned nucleosomes in yeast (49) and mammalian cells (Chapter 8 and refs 9, 50). A variation of LMPCR using ligation of the linker to double-stranded DNA prior to primer extension has been used with MNase to reveal nucleosomes over a liver-specific enhancer. McPherson *et al.* observed that three precisely positioned nucleosomes were found only in liver, where the enhancer is active (50). DMS is a very small alkylating agent and does not reveal nucleosomes; however, it is likely that more bulky alkylating agents will be more strongly affected by chromatin structure. Larger alkylating agents have recently been used for LMPCR (51, 52) but nucleosomes were not studied. Another recent study by Wei *et al.* (53) used the *E. coli* UvrABC excinuclease complex and LMPCR to study the repair of benzo[a]pyrene diol epoxide adducts. In theory, the UvrABC excinuclease should enable the use of LMPCR in conjunction with psoralen adducts, although this has not yet been reported. Another interesting approach to the study of agents that crosslink DNA and/or otherwise create road blocks to the progression of polymerases engaged in primer extension is to analyse the new molecules produced by the primer extension reaction. These molecules, which end in a 3′ hydroxyl, are normally not analysed by the LMPCR procedure. All information comes from the 'old' template strand. However, Grimaldi *et al.* (54) have reported that RNA ligase can be used to add an DNA linker to the 3′ end of primer-extended molecules; exponential PCR can then be done, as for LMPCR. Although promising, the sensitivity of this method needs to be improved. If sensitivity can be increased, this procedure has the potential to reveal features of DNA, such as psoralen crosslinks and other adducts, that will block the progression of the polymerase during primer extension. With regard to sensitivity, Mirkovitz *et al.* (55, 56) have developed a method for genomic sequencing that depends on enriching for the target sequence by hybridization to an excess of *in vitro* synthesized, biotin-labelled RNA and collecting the hybrid molecules. This enrichment method should be easily adaptable for use with LMPCR, but has not yet been reported.

One minor drawback of the LMPCR procedures as usually performed is that moderate amounts of radioactivity are needed. However, the signal strength seen for most LMPCR experiments is such that non-radioactive detection should be adequate. Preliminary experiments have, in fact, shown that non-radioactive detection is possible, either by hybridization with a biotinylated probe, or by the use of a 5′ fluorescent-labelled third primer. So far, however, the signal-to-noise ratio is not as good as for radioactive detection (unpublished). Halle *et al.* (13) have reported a method for the direct transfer of LMPCR fragments to a moving membrane during electrophoresis, and then non-radioactive detection by hybridization with biotin-labelled probes. Although expensive special equipment is needed for this approach, it is clear that non-radioactive detection and partial automation of LMPCR have good future potential.

References

1. Galas, D. J. and Schmitz, A. (1978). *Nucl. Acids Res.*, **5**, 3157.
2. Church, G. M., Ephrussi, A., Gilbert, W., and Tonegawa, S. (1985). *Nature*, **313**, 798.
3. Ephrussi, A., Church, G. M., Tonegawa, S., and Gilbert, W. (1985). *Science*, **227**, 134.
4. Becker, P. B., Ruppert, S., and Schütz, G. (1987). *Cell*, **51**, 435.
5. Saluz, H. P. and Jost, J. P. (ed.) (1987). *A laboratory guide to genomic sequencing*. Birkhauser Verlag, Basel.
6. Mueller, P. R. and Wold, B. (1989). *Science*, **246**, 780.
7. Pfeifer, G. P., Steigerwald, S. D., Mueller, P. R., Wold, B., and Riggs, A. D. (1989). *Science*, **246**, 810.
8. Pfeifer, G. P., Tanguay, R. L., Steigerwald, S. D., and Riggs, A. D. (1990). *Genes Dev.*, **4**, 1277.
9. Pfeifer, G. P., and Riggs, A. D. (1991). *Genes Dev.*, **5**, 1102.
10. Rozek, D. and Pfeifer, G. (1993). *Mol. Cell Biol.*, **13**, 5490.
11. Pfeifer, G. P., Drouin, R., Riggs, A. D., and Holmquist, G. P. (1992). *Mol. Cell. Biol.*, **12**, 1798.
12. Tornaletti, S. and Pfeifer, G. P. (1995). *J. Mol. Biol.*, **249**, 714.
13. Halle, J. P., Wurst, H., and Schmidt, C. (1993). *DNA Seq.*, **3**, 283.

14. Riggs, A. D. and Pfeifer, G. P. (1997). PCR-aided genomic footprinting. In *Advances in Mol. and Cell Biol.* Series title: *In vivo footprinting* (ed. I. L. Cartwright), Vol. 21, pp. 47–72 JAI Press, Greenwich, Conneticut.
15. Kara, C. J. and Glimcher, L. H. (1993). *EMBO J.*, **12**, 187.
16. Weih, F., Nitsch, D., Reik, A., Schütz, G., and Becker, P. B. (1991). *EMBO J.*, **10**, 2559.
17. Faber, S., Ip, T., Granner, D., and Chalkley, R. (1991). *Nucl. Acids Res.*, **19**, 4681.
18. Wu, L. and Whitlock Jr., J. P. (1992). *Proc. Natl Acad. Sci. USA*, **89**, 4811.
19. Toth, M., Doerfler, W., and Shenk, T. (1992). *Nucl. Acids Res.*, **20**, 5143.
20. Riggs, A. D. and Pfeifer, G. P. (1992). *Trends Genet.*, **8**, 169.
21. Wolffe, A. P. (1994). *Dev. Genet.*, **15**, 463.
22. Schlossherr, J., Eggert, H., Paro, R., Cremer, S., and Jack, R. S. (1994). *Mol. Gen. Genet.*, **243**, 453.
23. Steigerwald, S. D., Pfeifer, G. P., and Riggs, A. D. (1990). *Nucl. Acids Res.*, **18**, 1435.
24. Tommasi, S., LeBon, J. M., Riggs, A. D., and Singer-Sam, J. (1993). *Somat. Cell Mol. Genet.*, **19**, 529.
25. Wijnholds, J., Philipsen, J. N. J., and Ab, G. (1988). *EMBO J.*, **7**, 2757.
26. Glasel, J. A. (1995). *Biotechniques (United States)*, **18**, 62.
27. Sambrook, J., Fritsch, E. F., and Maniatis, T. (ed.) (1989). *Molecular cloning: a laboratory manual* (2nd edn). Cold Spring Harbor Press, Cold Spring Harbor, New York.
28. McGhee, J. D. and Felsenfeld, G. (1979). *Proc. Natl Acad. Sci. USA*, **76**, 2133.
29. Kochanek, S., Renz, D., and Doerfler, W. (1993). *Nucl. Acids Res.*, **21**, 5843.
30. Drew, H. R. and Travers, A. A. (1984). *Cell*, **37**, 491.
31. Miller, M. R., Castellot, J. J., and Pardee, A. B. (1978). *Biochemistry*, **17**, 1073.
32. Zhang, L. and Gralla, J. D. (1989). *Genes Dev.*, **3**, 1814.
33. Hornstra, I. K. and Yang, T. P. (1993). *Anal. Biochem.*, **213**, 179.
34. Singer, J., Roberts-Ems, J., and Riggs, A. D. (1979). *Science*, **203**, 1019.
35. Tormanen, V. T., Swiderski, P. M., Kaplan, B. E., Pfeifer, G. P., and Riggs, A. D. (1992). *Nucl. Acids Res.*, **20**, 5487.
36. Pfeifer, G. P., Singer-Sam, J., and Riggs, A. D. (1993). In *Methods in enzymology* (ed. J. Abelson and M. Simon), Vol. 225, pp. 567–83. Academic Press, New York.
37. Maxam, A. M. and Gilbert, W. (1980). In *Methods in enzymology* (ed. J. Abelson and M. Simon), Vol. 65, pp. 499–560. Academic Press, New York.
38. Iverson, B. L. and Dervan, P. B. (1987). *Nucl. Acids Res.*, **15**, 7823.
39. Smit, A. F. A., Toth, G., Riggs, A. D., and Jurka, J. (1995). *J. Mol. Biol.*, **246**, 401.
40. Smit, A. F. A. and Riggs, A. D. (1995). *Nucl. Acids Res.*, **23**, 98.
41. Rychlik, W. and Rhoads, R. E. (1989). *Nucl. Acids Res.*, **17**, 8543.
42. Church, G. M. and Kieffer-Higgens, S. (1988). *Science*, **240**, 185.
43. Pfeifer, G. P. and Riggs, A. D. (1993). In *Methods in molecular biology: PCR protocols: current methods and applications* (ed. H. Griffin and A. Griffin), Vol. 15, pp. 153–68. Humana Press, Totowa, NJ.
44. Garrity, P. A. and Wold, B. J. (1992). *Proc. Natl Acad. Sci. USA*, **89**, 1021.
45. Quivy, J. P. and Becker, P. B. (1993). *Nucl. Acids Res.*, **21**, 2779.
46. Singer-Sam, J., Robinson, M. O., Bellvé, A. R., Simon, M. I., and Riggs, A. D. (1990). *Nucl. Acids Res.*, **18**, 1255.
47. Jenssen-Eller, K., Picozza, E., and Crivello, J. F. (1994). *Biotechniques*, **17**, 962.

48. Mirkovitch, J. and Darnell Jr., J. E. (1992). *Mol. Biol. Cell*, **3**, 1085.
49. Fisher-Adams, G. and Grunstein, M. (1995). *EMBO J.*, **14**, 1468.
50. McPherson, C., Shim, E., Friedman, D., and Zaret, K. (1993). *Cell*, **75**, 387.
51. Lee, C., Pfeifer, G., and Gibson, N. (1994). *Biochemistry*, **33**, 6024.
52. Lee, C., Pfeifer, G., and Gibson, N. (1994). *Cancer Res.*, **54**, 1622.
53. Wei, D., Maher, V. M., and McCormick, J. J. (1995). *Proc. Natl Acad. Sci. USA*, **92**, 2204.
54. Grimaldi, K. A., McAdam, S. R., Souhami, R. L., and Hartley, J. A. (1994). *Nucl. Acids Res*, **22**, 2311.
55. Mirkovitch, J. and Darnell Jr., J. E. (1991). *Genes Dev.*, **5**, 83.
56. Mirkovitch, J., Decker, T., and Darnell Jr., J. E. (1992). *Mol. Cell. Biol.*, **12**, 1.

5

Analysis of nuclear scaffold attachment regions

SUSAN M. GASSER and YEGOR S. VASSETZKY

1. Introduction

1.1 Definitions

The eukaryotic nucleus is a complex, highly organized structure. Separation into soluble and insoluble subfractions is one of the few ways to analyse nuclear organization. As long as 20 years ago, it was observed that the removal of histones by high salt extraction resulted in a halo of tortionally constrained loops of naked DNA, emanating from a residual nuclear structure (1–5). The halo size changed on addition of ethidium bromide, demonstrating that the individual loops could become supercoiled, and are thus topologically independent domains.

The proteinaceous structures that remained after membrane- and histone-extraction have been called scaffolds (4, 6), nuclear matrices (7), cages (8), nucleo- or karyoskeletons (9, 10), or the nuclear matrix-pore, complex-lamina fraction (NMPCL; 11). As described below, the composition of these various substructures reflects the protocol used for nuclear isolation and for histone depletion. Only the nuclear lamina meshwork and the pore complexes were invariably associated with the residual nuclear material after extraction. The lamina and pore network also clearly exist as structural elements in intact cells. However, the recovery of the internal, ill-defined network of fibres, apparently containing both protein and RNA, and the recovery of residual nucleolar material, depends largely on the conditions of extraction. Since the various nuclear subfractions are operationally defined, we will first describe what we mean by each of these terms, and then discuss briefly how various parameters of isolation influence their composition.

The nuclear *matrix* generally refers to a structure obtained from interphase nuclei that is resistant to extraction by non-ionic detergent and high concentrations of NaCl. One major point of variation in different preparations of nuclear matrices is the introduction of a nuclease digestion step performed in the presence of Mg^{2+} or Ca^{2+} at 37°C, prior to histone extraction. This digestion step can induce many otherwise soluble proteins to become insoluble, and constitutes what is euphemistically called a 'stabilization' of nuclear structure (see below). To distinguish between more and less complex matrices, we will use terminology introduced by Lebkowski and Laemmli (12, 13). That is, type I nuclear matrix results from the extraction of histones with 1 or 2 M NaCl from nuclei that have been exposed, even if briefly, to either Cu^{2+}, Ca^{2+} at 37 °C, or combinations of divalent cations, while type II matrices have never been exposed to oxidizing conditions, heavy metals, or temperatures above 20 °C prior to histone extraction. Morphologically, type II matrices appear as empty spheres, whose shape is defined by the nuclear lamins and pore complex proteins, which together form a lattice just under the nuclear membrane. Type I matrices have a much more complex protein pattern and contain internal nuclear material which has a fibrous appearance in electron microscopic images (13–15). The presence of a residual nucleolar structure depends on whether or not the nuclei were treated with RNase prior to extraction (10, 16, 17).

If no nuclease digestion occurs prior to histone extraction, the genomic DNA is constrained in type II matrices as a large, slowly sedimenting halo of DNA apparently attached to the nuclear lamina. The DNA halo of type I matrices is significantly smaller (13), apparently reflecting additional DNA interactions with internal nuclear material. It is not known to what degree the loops of DNA represent domains that exist in intact nuclei or result from the extraction procedure.

To optimize the isolation of DNA fragments bound to the residual nuclear structure a protocol was developed in Laemmli's laboratory, which entailed stabilization of nuclei by a brief exposure either to Cu^{2+} or 37 °C, followed by extraction of histones by lithium 3′,5′-diiodosalicylate (LIS; 18). LIS is a detergent-like salt which efficiently solubilizes histones and certain non-histone proteins at very low concentrations in either low or physiological salt buffers. The resulting nuclear *scaffold* has a complex protein pattern, like the nuclear matrix type I, but is able to maintain preferential interaction with dispersed AT-rich sequences in the genome. These were called SARs for scaffold attachment regions, and could be mapped by restriction enzyme digestion of extracted nuclei, and comparative Southern blots of bound and released DNA.

1.2 Parameters affecting scaffold and matrix composition

As mentioned above, the composition and structural stability of the karyoskeleton is profoundly affected by a brief incubation of nuclei at moderately elevated temperatures *in vitro* (12, 13, 19, 20), or by heat-shock conditions *in vivo* (20, 21). For unknown reasons, exposure to either Cu^{2+} or Ag^{2+} also 'stabilizes' nuclear proteins, mimicking a step initially used in the isolation of metaphase chromosomal scaffolds (22). In metaphase chromosomes the effects of copper are reversible by 2-mercaptoethanol or strong copper chelators. This reversal was argued to reflect metal depletion and not sulfhydryl reduction, since $NaBH_4$, a strong reducing agent, had no effect on the stabilization of scaffolds (22). It is possible that certain non-histone chromosomal proteins bind Cu^{2+} very tightly, and that the presence of the ligand alters their solubility.

A third efficient means to 'stabilize' scaffolds is the crosslinking of the sulfhydryl groups of nuclear proteins with sodium tetrathionate (23). The effects of thiol crosslinking and/or alkylating reagents on the composition of the nuclear matrix has been reviewed extensively by Kaufmann *et al.* (24), and will not be discussed further here. Since the protein composition of nuclear scaffolds and matrices varies greatly depending on the conditions used for stabilization and extraction, one is inevitably led to question the importance of finding a given protein in this fraction. In contrast, there are only minor differences in the patterns of DNA fragments recovered with the nuclear scaffold when copper and heat techniques were compared in the mapping of scaffold-attached DNA (18). More significant variation is observed when different means are used to extract histones.

1.3 Conditions of histone extraction

Polyanions (like heparin mixed with dextran sulfate), high concentrations of salt, or low concentrations of lithium 3′,5′-diiodosalicylate (LIS) are most commonly used to remove histones from nuclei and chromosomes. Originally, polyanions were used to remove histones from HeLa metaphase chromosomes (4, 25). The resulting structures were stable in gradients containing 2 M NaCl, which had been shown previously to dissociate histones from DNA (26). In the case of interphase nuclei, high salt concentrations efficiently remove histones, but appear to trap (or maintain) transcription complexes in a nuclear matrix fraction, while extraction with lithium 3′,5′-diiodosalicylate (LIS) does not (18, 27, 28). This detergent-like reagent had been used previously in the isolation of glycoproteins from cell membranes (29), and is extremely efficient at 'salting in' or solubilizing proteins at concentrations under 20 mM (30). LIS is usually used in conjunction with a low level of a non-ionic detergent, such as digitonin, which enhances both the solubility of LIS itself and the efficiency of extraction of membrane proteins and histones.

Another major difference between the material released by LIS compared with high salt is the release of nuclear RNA and what appears to be splicing complexes. If no RNase treatment is included prior to extraction, RNA remains as a major component in the salt-extracted matrix of mammalian and *Drosophila* nuclei (31), while most RNA is released by extraction with LIS, at least in *Drosophila* and yeast.

1.4 Mapping DNA bound to the nuclear scaffold

The study of nuclear scaffolds and matrices has identified a structurally related class of DNA sequences that interact *in vitro* with this structure. Two techniques have been used to map scaffold- or matrix-attached DNA. In the first, genomic DNA must be maintained intact until after histone extraction. Then residual nuclei and their attached haloes of DNA are washed, digested by restriction enzymes, and the DNA fragments bound to the scaffold (SARs) are sedimented by low-speed centrifugation. Southern blot analysis of bound and released DNA usually reveals a different pattern of fragments for bound- and released-fractions. This approach works very well for organisms with fairly small genomes, such as yeast and *Drosophila* (18, 32, 33). With the larger genomes, which produce larger haloes, DNA was found to shear during manipulation. For this reason, a second approach was introduced by Garrard and colleagues for working with mouse L cell nuclei (34). Salt-extracted nuclear matrices from which most nuclear DNA was removed by DNase I digestion were prepared. End-labelled fragments of cloned DNA were then added back to the matrices in the presence of a non-specific competitor, to define sequences (MARs) that bound preferentially to nuclear matrix proteins. In most cases, scaffold attachment regions defined by the first method are also able to bind to isolated matrices. The opposite is not always true.

The scaffold- or matrix-associated regions (SARs or MARs) revealed striking overall sequence similarity, although little primary sequence identity. In general, they are found in non-coding regions, they are >70% A+T, longer than 200 bp in length, and contain sequence patterns that form DNA with a characteristic narrow minor groove, a structure which results from the stacking of sequential thymidine residues (35, 36). Drugs such as distamycin, that interact specifically with this minor groove structure, compete efficiently for the SAR-scaffold binding (37). The size requirement for SAR-scaffold interaction may reflect multiple sites of weak interaction, involving either structure-specific or sequence-specific recognition. Several abundant DNA-binding proteins, like topoisomerases I and II, histone H1, and HMG I/Y proteins, bind preferentially, yet still with relatively low affinity, to DNA having a narrow minor groove. Consistently, topoisomerases and histone H1 were shown to bind cooperatively to SAR DNA *in vitro*, coating the DNA strand and provoking its precipitation (38, 39).

The function of SAR elements in intact cells is still not clear. Recent studies demonstrated that the SARs enhance transcription of DNA sequences randomly integrated into genome, although probably not by acting as barrier elements that protect the gene from surrounding heterochromatin (40, 41). The unwinding capacity of many SAR elements suggests they may serve to destabilize the double helix at replication origins (42), while the preferential staining of A+T-rich and G+C-rich regions in condensed metaphase chromosomes is consistent with a role for SARs in higher-order DNA organization (43). A likely possibility is that these regions facilitate the binding of topology-modulating proteins like histone H1, topoisomerases, and HMG proteins, as discussed (44).

Here we will describe both a high-salt and a LIS method for extraction of the nuclei and will discuss some of the differences with other available methods. The protocols are modified from published work (18, 32, 33, 45). It is extremely important to keep in mind that nothing that is shown to be true for matrices or scaffolds is necessarily true for intact nuclei. Thus such studies, no matter how carefully done, can only give a hint as to how nuclei might be organized. As for most biochemical studies, genetic evidence defining a function *in vivo*, is required to confirm hypotheses advanced from examination of the nuclear matrix or scaffold.

2. Isolation of nuclei

Nuclear isolation methods must be adapted for different species and cell types (see Chapters 1–3 for more details). We describe, here, the isolation of interphase nuclei from HeLa S3 cells grown in suspension. This method can be adapted to *Drosophila* cultured cells with minor variations (but see Chapter 3). The buffers are based on the EDTA-containing spermine/spermidine buffers of Hewish and Burgoyne (46), which are essential for keeping genomic DNA fully intact (cf. Chapter 2, *Protocol 2*). Cell disruption is performed in the presence of 0.1% digitonin in a hypotonic solution using a Dounce homogenizer. Either Ficoll (Sigma) or sucrose can be added if osmotic stabilizers are desired. Percoll (Pharmacia) gradients can efficiently separate nuclei from cell and cytoplasmic debris, although it may not be necessary for many types of experiment.

Protocol 1. Isolation of nuclei

Equipment and reagents

Make the following fresh from stock solutions.

- MEM-S medium (Gibco) supplemented with 10% fetal calf serum
- 10 × A (may be stored at 4°C): 200 mM Tris–HCl, pH 7.4, 20 mM EDTA–KOH, pH 7.5, 5 mM spermidine, 2 mM spermine, 800 mM KCl
- Digitonin (stored frozen as a 10% stock (w/v) in water)[a]
- Thiodiglycol (Pierce), an antioxidant in 25% or 100% solution
- Trasylol (Bayer), a non-toxic peptide inhibitor for serine proteases
- PMSF: 0.2 M stock in 100% ethanol. Store at – 20°C.
- Breakage buffer: 0.5 × A , 0.1% digitonin, 1% thiodiglycol, 0.5% Trasylol, 0.5 mM PMSF (Add digitonin and PMSF just prior to use.)
- Percoll (Pharmacia)
- 1% SDS
- Sorvall GS-3 rotor and HB-4 swing-out rotor
- Tight-fitting Dounce homogenizer

Method

1. Grow 0.5 litre of HeLa S3 cells in suspension to a density of approximately 2–5 × 10^5 cells/ml in MEM-S medium (Gibco) supplemented with 10% fetal calf serum.

Protocol 1. *Continued*

2. Harvest cells by centrifugation in a Sorvall GS-3 rotor at 1000 *g* at room temperature.
3. Wash the collected cells twice at room temperature by gentle resuspension (40 ml per 500 ml of culture), in 0.25 × A made by a 40-fold dilution of the 10 × A stock solution and containing 1% thiodiglycol, 0.5% Trasylol, 0.5 mM PMSF. Sediment the cells gently by centrifugation in a Sorvall GS-3 rotor at 400 *g* at room temperature. All the subsequent steps are performed at 4 °C.
4. Transfer the cells into 20 ml of chilled breakage buffer per 500 ml of cell suspension. Homogenize cells on ice with 20 vigorous strokes of a tight-fitting Dounce homogenizer.
5. Verify the efficiency of cell breakage using a phase-contrast microscope.
6. Spin the disrupted cells at 500 *g* for 10 min at 4°C. Repeat steps 4 through 6 at least once. Check by phase-contrast microscopy that the sedimented population is mostly nuclei.
7. Resuspend the pellet containing nuclei and residual cell debris in 35 ml of the 0.25 × A buffer with 60% (v/v) Percoll, 2 mM spermidine, 0.8 mM spermine, 0.1% digitonin, 0.5% thiodiglycol, 0.1% Trasylol, and 0.2 mM PMSF. Homogenize the mixture 10 times in a tight-fitting Dounce homogenizer and spin at 28 000 g for 20 min at 4 °C[b].
8. Recover clean nuclei as a white band near the bottom of the Percoll gradient. To remove Percoll from the recovered fraction of nuclei, dilute the mixture approximately threefold with 0.25 × A buffer with all additions as Step 7, but Percoll, and spin for 10 min at 1500 *g* in a Sorvall HB-4 swing-out rotor at 4 °C. Repeat the wash.
9. Measure the A_{260} of the nuclear suspension diluted 1/100 in 1% SDS.
10. Use the nuclei directly or store them at – 20 °C in the same buffer containing 50% glycerol.

[a] Digitonin as purchased can be very impure. Often to achieve a 10% solution (w/v) in water, it must first be recrystallized out of a saturated solution in methanol. To do this, dissolve 20 g digitonin in 50 ml 100% methanol at 55 °C, and filter. Let cool, recover recrystallized digitonin, dry under vacuum, weigh, and dissolve in water. Handle with mask and gloves.
[b] This step allows separation of nuclei from cytoskeletal debris, which usually stays at the top of the Percoll gradient.

The advantage of this method is that spermine/spermidine buffers containing EDTA minimize nucleolytic damage during nuclear isolation. Percoll-gradient purified nuclei are not necessary for scaffold isolation. However, cell debris interferes enormously with the extraction procedure, since it becomes entangled in the halo of DNA. Nuclei should be at least 90% pure by visual inspection before proceeding further. This method is not appropriate for animal tissues (see Chapter 1, *Protocol 1*, and ref. 45, for rat liver nuclei), and minor modifications are required for yeast (33) and *Drosophila* (Chapter 3, *Protocol 1*, and ref. 18).

3. Isolation of agarose-embedded cells and nuclei

Most procedures for the isolation of nuclei involve treatment of the nuclei in non-physiological conditions (hypotonic or hypertonic) since they tend to aggregate at physiological salt concentration. A method has been developed that avoids this by embedding cells in agarose beads and then extracting nuclei using various methods (47, 48). The pores in agarose beads are large enough to allow elution of macromolecules as large as 10^4 kDa, while intact chromosomal DNA remains within the beads. All the extraction steps can then be carried out at physiological salt concentration.

Protocol 2. Preparation of agarose-embedded cells

Reagents

- Phosphate-buffered saline (PBS)
- 2.5% agarose (Sigma type VII) in PBS
- Liquid paraffin

Method

1. Grow HeLa cells as in *Protocol 1*.
2. Melt 2.5% agarose, cool to 39 °C, and mix 1 vol. of agarose in a round-bottom flask with 4 vol. of cells in PBS at 39 °C.
3. Add 2 vol. of liquid paraffin at 39 °C to 1 vol. of cells in molten agarose.
4. Shake the mixture at 800 cycles/min for 30 sec at 20 °C and cool in ice for 5 min.
5. Sediment the beads in a bench-top centrifuge, 500–1000 r.p.m. or roughly 1000 *g*.
6. Remove the paraffin and excess aqueous phase.

The encapsulated cells can then be lysed, and their nuclear DNA extracted using one of the methods described below. For nuclei with large haloes this avoids the problem of shearing the genomic DNA. Isolated nuclei can also be embedded in agarose beads in spermine/spermidine buffers as described above. Histone extractions can also be carried out on nuclei in agarose beads, except that washes should be extended slightly in time. Note that cells are essentially heat-shocked by the embedding technique.

4. Isolation of nuclear scaffolds

Several methods exist for the isolation of nuclear scaffolds. If the aim is to isolate scaffold proteins alone, nuclei can be washed in a standard buffer without EDTA, digested with endonucleases at either 0 °C or 37 °C, for type I or type II matrices (see section 1), respectively, and then extracted with a solution containing either 10 mM LIS or 2 M NaCl, as described below. If the DNA fragments bound to the nuclear scaffolds are to be studied, extraction with LIS or NaCl must precede restriction enzyme digestion. Below we describe both LIS- and salt-extraction procedures.

The LIS-extraction procedure is widely used on *Drosophila*, mouse, and yeast cell nuclei to study the scaffold-bound DNA fragments (18, 32–34, 49–51). The optimal amount of LIS depends on the species used, but 10 mM works for *Drosophila*, yeast, and rat liver (e.g. ref. 45). We stress again that the nuclear scaffold/matrix is an intermediate product of disassembly of the cells nucleus, and the relationship between the situation observed in nuclear scaffolds/matrices and that in nuclei is not always direct.

Protocol 3. LIS extraction of nuclei

Reagents

- 10 × A^{-EDTA} (may be stored at 4 °C): 200 mM Tris–HCl, pH 7.4, 5 mM spermidine, 2 mM spermine, 800 mM KCl
- Digestion buffer: 20 mM Tris–HCl, pH 7.4, 0.05 mM spermine, 0.125 mM spermidine, 20 mM KCl, 70 mM NaCl, 10 mM $MgCl_2$, 0.1% digitonin, 1% Trasylol, 0.1% thiodiglycol, 0.1 mM PMSF
- Trasylol (Bayer)
- Thiodiglycol (Pierce)
- PMSF
- Proteinase K
- Low-salt extraction buffer: 5–20 mM LIS (Sigma), 20 mM Hepes–NaOH, pH 7.4, 100 mM Li-acetate, 0.1% digitonin, 2 mM EDTA–NaOH, pH 7.5. Prepare this by dissolving the desired amount of LIS in water, then adding digitonin, and a 5 × concentrated stock of Hepes, Li-acetate, and EDTA. Prepare just before use, and keep at room temperature. 10 mM LIS is a standard concentration to use.
- Appropriate restriction endonucleases

Method

1. Resuspend 10 A_{260} units of nuclei well, such that there are no clumps of nuclei, in 0.2 ml 0.25 × A buffer containing no EDTA with the standard additions (see *Protocol 1*). Incubate the suspension at 37 °C for 20 min. This can be raised to 42 °C for mammalian nuclei[a] (see ref. 45).
2. Return to ice or room temperature, and add, very slowly, 5 ml of extraction buffer containing 10 mM LIS. Slowly invert the tube several times during and after addition to mix evenly. Incubate for 5 min at room temperature[b].
3. Recover histone-depleted nuclei by centrifugation at 2500 *g* in a Sorvall HB-4 rotor for 20 min at room temperature. The nuclear haloes should form a fluffy white pellet.

4. Wash the pellet by adding 10 or 15 ml of digestion buffer (20 mM Tris-HCl, pH 7.4, 0.05 mM spermine, 0.125 mM spermidine, 20 mM KCl, 70 mM NaCl, 10 mM $MgCl_2$, 0.1% digitonin, 1% Trasylol, 0.1% thiodiglycol, 0.1 mM PMSF), and gently inverting the tube during 5 minutes incubation at room temperature[c]. Sediment by centrifugation for 10 min at 2000 *g*, either at room temperature or 4°C. If nuclease contamination is a problem, wash at 4°C and omit $MgCl_2$. Repeat the wash three or four times.
5. After the last centrifugation, gently shake the haloes off the bottom of the tube into a minimal volume of digestion buffer (e.g. 200 μl). Digest with appropriate restriction endonucleases at 20 U/A_{260} nuclei for 3 h at 37°C or 100 μg/ml DNase I for 1 h at 4°C (supplement the buffer with 1 mM $CaCl_2$ for DNase I digestion). Add more restriction enzymes after 1 h.
6. Separate digested DNA from the scaffolds by diluting to 500 μl final volume with digestion buffer, and centrifuging at 2500 *g* for 10 min at 4°C. Save the supernatant. Add another 300 μl digestion buffer, vortex or pipette well with an Eppendorf pipette, and repeat the centrifugation[d].
7. Extract the scaffold-bound DNA by adding 40 mM EDTA, 1% SDS, and 1 mg/ml proteinase K. Combine the supernatants containing solubilized DNA and adjust to 0.2% SDS, 40 mM EDTA, 0.2 mg/ml proteinase K. Incubate the protease digestions either overnight at 37°C or for 3 h at 56°C. Obtain DNA by standard phenol chloroform extraction and ethanol precipitation, and load in equal amounts for pellet and supernatant fractions on an agarose gel. Probe for appropriate fragments by Southern hybridization[e].

[a] This 'stabilizing step' can be replaced by incubating with 1 mM $CuSO_4$ for 10 min at 4°C prior to histone extraction.
[b] This is a critical point in the preparation. Especially with large genomes, if the nuclei lyse, the scaffolds cannot be sedimented and washed in the following steps. If this happens try again with less LIS. Use a 15 ml, round-bottomed, polycarbonate tube to avoid extracted nuclei sticking to the tube. Glass should be avoided.
[c] The nuclear haloes will float back and forth in the tube.
[d] The scaffolds, which are recovered in the pellet, can be analysed for both protein and DNA. For protein analysis the fraction is simply denatured in 2% SDS.
[e] DNA can also be redigested with restriction enzymes after purification, to eliminate partially digested fragments.

Protocol 4. High-salt method of isolation of nuclear scaffold

The high-salt extraction method is widely used in studies of the nuclear scaffold (7, 12, 13, 34). In some cases, results are similar to those obtained using LIS extraction, with one major exception: the scaffolds obtained by 2 M NaCl extraction are enriched in transcribed sequences, which makes the analysis of the DNA fragments attached to the nuclear scaffolds more complicated, since it reflects the transcriptional state of the cell.

Protocol 4. *Continued*

Reagents

- Buffer 1: 0.25 × A buffer (*Protocol 1* without EDTA) plus 10 mM $MgCl_2$, 50 mM NaCl, 0.1% digitonin, 0.5% thiodiglycol, 0.1% Trasylol, 0.2 mM PMSF
- SDS
- 2 × high salt extraction buffer: 4 M NaCl, 40 mM Tris–HCl, pH 7.4, 20 mM EDTA–NaOH
- Appropriate restriction endonucleases or DNase I

Method

1. Resuspend 5 A_{260} units of nuclei in 0.5 ml buffer 1.
2. Digest with appropriate restriction endonucleases at 20 U/ A_{260} for 3 h at 37 °C or 100 μg/ml DNase I for 1 h at 4 °C.
3. If incubated at 4 °C, then add $CuSO_4$ to 1 mM, leave for 10 min at 4 °C.
4. Add an equal volume of 2 × high-salt extraction buffer. Incubate for 20 min at 4 °C.
5. Recover the extracted nuclei by centrifugation at 2000 *g* for 10 min at 4 °C. Save the supernatant.
6. Wash the pellet three times in 1 ml of 1 × extraction buffer (2 M NaCl, 20 mM Tris–HCl, pH 7.4, and 10 mM EDTA–NaOH).
7. Add SDS to 2% and denature by heating to 95 °C for protein analysis[a].

[a] In our hands, this is not a good preparation for mapping scaffold associated fragments by Southern hybridization, but it works in the rebinding assay described in Protocol 5.

5. Analysis of DNA associated with the nuclear scaffold

DNA associated with the nuclear scaffolds may be studied in two ways: it can be extracted from the scaffolds and recovered from the supernatant and analysed by gel electrophoresis, Southern blotting, and hybridization with the probes of interest. Another method involves monitoring the binding of end-labelled fragments to nuclear scaffolds *in vitro*, described below. Large regions of the genome may have to be tested to find fragments that bind with high affinity to scaffolds.

Protocol 5. *In vitro* scaffold binding assay

In this method the DNA of interest is cleaved by restriction enzymes to fragments ranging from 200 bp to 2 kb in length. They are radioactively labelled and then incubated with isolated nuclear scaffolds in the presence of competitor DNA or the endogenous genomic DNA. Fragments that have an affinity for the nuclear scaffold bind, while others remain in solution. Comparison of the spectrum of input and bound DNA fragments allows one to estimate the potential of a given DNA sequence for interaction with the nuclear scaffold. Please note that relative affinity should be determined by titrating competitor DNA.

Equipment and reagents

- Digestion buffer: 20 mM Tris–HCl, pH 7.4, 0.05 mM spermine, 0.125 mM spermidine, 20 mM KCl, 70 mM NaCl, 10 mM $MgCl_2$, 0.1% digitonin, 1% Trasylol, 0.1% thiodiglycol, 0.1 mM PMSF
- 20 mM EDTA, 20 mM Tris–HCl, pH 7.4, 0.1 M NaCl
- SDS
- Proteinase K
- Phenol
- DEAE 81 paper (Whatman)
- Autoradiographic film
- Agarose gels

Method

1. After incubating histone-depleted nuclei with restriction endonucleases (*Protocol 3*, step 5), add the equivalent of between 0.2 and 0.5 A_{260} units of starting nuclei to the end-labelled probe (2–10 ng, or about 10^5 c.p.m.) representing the domain of interest in the form of a fragment ranging from 200 bp to 2 kb in length[a]. Alternatively, use DNase-digested or salt-extracted matrices in 100 μl of the digestion buffer with 20 mM EDTA, up to 20 ng of the end-labelled DNA of interest, and 1–10 μg of non-specific competitor, such as sonicated total *E. coli* DNA[b]. Incubate for 1 h or more at 37 °C, shaking gently.
2. After incubation of the scaffolds, dilute the mixture slightly with digestion buffer and recover scaffolds with the bound DNA by centrifugation at 3000 *g* for 10 min at 4 °C.
3. Save the supernatant, and wash the scaffold fraction three times with 1 ml chilled digestion buffer. Spin at 5000 *g* for 10 min at 4 °C for each wash.
4. Resuspend the scaffold pellet in 20 mM EDTA, 20 mM Tris–HCl, pH 7.4, and 0.1 M NaCl. Add SDS to 1% and proteinase K to 100 μg/ml and digest for 1–3 h at 56 °C. Adjust the supernatant to 20 mM EDTA, and also treat with proteinase K.
5. Extract with phenol chloroform.
6. Precipitate DNA with ethanol.
7. Use directly for electrophoresis in agarose gels. Include the input DNA, the scaffold-bound DNA, and the DNA that did not bind to the scaffolds. Dry the gel on DEAE 81 paper and expose to autoradiographic film.

[a] If the end-labelled probe is digested with the same enzymes used to digest the extracted nuclei, the probe can be added during the last hour or two of digestion, and the genomic DNA serves as competitor DNA.

[b] Exogenous competitor DNA must be added since there is no genomic DNA as competitor.

The relative amounts of the probe, competitor DNA, and scaffolds are indicative, and must be adjusted for scaffolds prepared from different species and cell types. Adaptation of this protocol for budding and fission yeast is described in Amati and Gasser (33, 52) and adaptation to mitotic chromosomal scaffolds is described in Mirkovitch *et al.* (53). The relative ratio of competitor and probe can be standardized using a known scaffold-binding fragment, such as a MAR for the immunoglobulin genes (34) or various *Drosophila* SARs (18, 32, 49). The conditions that allow attachment of $>80\%$ of SAR DNA to the scaffold with $<10\%$ of non-SAR DNA attached can be considered satisfactory for the binding reactions.

6. Interpretation of the data obtained

Nuclear scaffolds are obtained from intact nuclei by extraction of histones and the removal of most DNA, so it is evident that their properties depend on the methods of extraction and reagents used. The identification of a specific region as scaffold-binding is highly reproducible, and the fragments, as a rule, represent a family of related A+T-rich DNA sequences. Finding that a fragment has SAR activity is of interest only if put into a larger context. That is, its potential function in the chromosome should be explored using other techniques. One might ask the following questions. Does this fragment play a known role in DNA replication? Is it involved in higher-order chromatin organization? Is it associated with a tissue-specific enhancer, or a boundary element ? Are critical factors for opening or closing chromatin domains known to bind this region ?

It has been shown that isolation of matrices is useful for enriching in newly replicated DNA and for replication intermediates (54). Matrix- or scaffold-protein preparations might also therefore be enriched in proteins that organize origins of replication. Thus these subfractionation schemes are not only valuable for mapping mammalian origins of replication, but may also help in the isolation of relatively low abundance origin binding proteins.

References

1. Cook, P. R. and Brazell, I. A. (1975). *J. Cell Sci.*, **19**, 261.
2. Cook, P. R. and Brazell, I. A. (1976). *J. Cell Sci.*, **22**, 287.
3. Benyajati, C. and Worcel, A. (1976). *Cell*, **9**, 393.
4. Paulson, J. R. and Laemmli, U. K. (1977). *Cell*, **12**, 817.
5. Vogelstein, B., Pardoll, D. M., and Coffey, D. S. (1980). *Cell*, **22**, 79.
6. Adolph, K. W. (1980). *J. Cell Sci.*, **42**, 291.
7. Berezney, R. and Coffey, D. (1974). *Biochem. Biophys. Res. Commun.*, **60**, 1410.
8. Cook, P. R. and Brazell, I. A. (1980). *Nucl. Acids Res.*, **8**, 2895.
9. Miller, T. W., Huang, C.-Y., and Pogo, A. O. (1978). *J. Cell Biol.*, **76**, 675.
10. Hancock, R. (1982). *Biol. Cell*, **46**, 105.

11. Fisher, P. A., Berrios, M., and Blobel, G. (1982). *J. Cell. Biol.*, **92**, 674.
12. Lebkowski, J. S. and Laemmli, U. K. (1982). *J. Mol. Biol.*, **156**, 309.
13. Lebkowski, J. S. and Laemmli, U. K. (1982). *J. Mol. Biol.*, **156**, 325.
14. Hancock, R. and Hughes, M.E. (1982). *Biol. Cell*, **44**, 201.
15. He, D., Nickerson, J. A., and Penman, S. (1990). *J. Cell Biol.*, **110**, 569.
16. Bouvier, D., Hubert, J., Seve, A. P., and Bouteille, M. (1982). *Biol. Cell*, **43**, 143.
17. Lewis, C. D., Lebkowski, J. S., Daly, A. K., and Laemmli, U. K. (1984). *J. Cell Sci.*, **1** (suppl.), 103.
18. Mirkovitch, J., Mirault, M.-E., and Laemmli, U. K. (1984). *Cell*, **39**, 223.
19. Evan, G. I. and Hancock, D. C. (1985). *Cell*, **43**, 253
20. McConnell, M., Whalen, A. M., Smith, D. E., and Fisher, P. A. (1987). *J. Cell Biol.*, **105**, 1087.
21. Littlewood, T. D., Hancock, D. C., and Evan, G. I. (1987). *J. Cell Sci.*, **88**, 65.
22. Lewis, C. and Laemmli, U. K. (1982). *Cell*, **29**, 171.
23. Kaufmann, S. H. and Shaper, J. H. (1984). *Exp. Cell Res.*, **155**, 477.
24. Kaufmann, S. H., Fields, A. P., and Shaper, J. H. (1986). *Meth. Achiev. Exp. Pathol.*, **12**, 141.
25. Adolph, K. W., Cheng, S. M., and Laemmli, U. K. (1977). *Cell*, **12**, 805.
26. Spelsberg, T.L. and Hnilica, L.S. (1971). *Biochem. Biophys Acta*, **228**, 202.
27. Kirov, N., Djondjurov, L., and Tsanev, R. (1984). *J. Mol. Biol.*, **180**, 601.
28. Roberge, M., Dahmus, M. E., and Bradbury, E. M. (1988). *J. Mol. Biol.*, **201**, 545.
29. Marchesi, V. T. and Andrews, E. P. (1971). *Science*, **174**, 1247.
30. Robinson, D. R. and Jencks, N. P. (1965). *J. Am. Chem. Soc.*, **87**, 2470.
31. Nickerson, J. A., Krochmalnic, G., Wan, K. M., and Penman, S. (1989). *Proc. Natl Acad. Sci. USA*, **86**, 177.
32. Mirkovitch, J., Spierer, P., and Laemmli, U. K. (1986). *J. Mol. Biol.*, **190**, 255.
33. Amati, B. B. and Gasser, S. M. (1988). *Cell*, **54**, 967.
34. Cockerill, P. N. and Garrard, W. T. (1986). *Cell*, **44**, 273.
35. Yoon, C., Privé, G. G., Goodsell, D. S., and Dickerson, R. E. (1988). *Proc. Natl Acad. Sci. USA*, **85**, 6332.
36. Nelson, H. C. M., Finch, J. T., Luisi, B. F., and Klug, A. (1987). *Nature*, **330**, 221.
37. Kaes, E., Izaurralde, E., and Laemmli, U. K. (1989). *J. Mol. Biol.*, **210**, 587.
38. Izarraulde, E., Kaes, E., and Laemmli, U. K. (1989). *J. Mol. Biol.*, **210**, 573.
39. Adachi,Y., Kaes, E., and Laemmli, U. K. (1989). *EMBO J.*, **8**, 3997.
40. Phi-Van, L., von Kries, J. P., Ostertag, W., and Strätling, W. H. (1990). *Mol. Cell. Biol.*, **10**, 2302.
41. Poljak, L., Seum, C., Mattioni, T., and Laemmli, U. K. (1994). *Nucl. Acids Res.*, **22**, 4386.
42. Amati, B., Pick, L., Laroche, T., and Gasser, S. M. (1990). *EMBO J.*, **9**, 4007.
43. Saitoh, Y. and Laemmli, U. K. (1994). *Cell*, **76**, 609.
44. Laemmli, U. K., Kaes, E., Poljak, L., and Adachi, Y. (1992). *Curr. Opin. Genet. Dev.*, **2**, 275.
45. Izaurralde, E., Mirkovitch J., and Laemmli, U. K. (1988). *J. Mol. Biol.*, **200**, 111.
46. Hewish, D. R. and Burgoyne, L. A. (1973). *Biochem. Biophys. Res. Comm.*, **52**, 504.
47. Cook, P. R. (1984). *EMBO J.*, **3**, 1837.
48. Jackson, D.A. and Cook, P.R. (1985). *EMBO J.*, **4**, 919.
49. Gasser, S. M. and Laemmli, U. K. (1986). *Cell*, **46**, 521.

50. Gasser, S. M. and Laemmli, U. K. (1986). *EMBO J.*, **5**, 511.
51. Cardenas, M. E., Laroche, T., and Gasser, S. M. (1990). *J. Cell Sci.*, **96**, 439.
52. Amati, B. and Gasser, S. M. (1990). *Mol. Cell. Biol.*, **10**, 5442.
53. Mirkovitch, J., Gasser, S. M., and Laemmli, U. K. (1988). *J. Mol. Biol.*, **200**, 101.
54. Dijkwel, P. A., Vaughn, J. P., and Hamlin, J. L. (1991). *Mol. Cell. Biol.*, **11**, 3850.

6

Electron microscopy of chromatin

FRITZ THOMA

1. Introduction

Electron microscopy (EM) has been used to investigate how DNA is folded in nucleosomes, chromatin fibres, and higher-order structures. In contrast to biochemical and biophysical techniques, which produce averaged information of a whole population of structures, EM allows one to visualize and analyse individual molecules or chromatin complexes; averaged information must then come from the analysis of a significant number of complexes.

Over the years, various techniques have been exploited to visualize genomic chromatin in nuclei. Whole nuclei were lysed in low-salt and high-pH conditions and unfolded chromatin was visualized after centrifugation on a support grid. The results showed chromatin as 'beads-on-a-string' and provided electron microscopic evidence for nucleosomes as repetitive units of chromatin fibres (1). This technique was originally developed to maximally spread out and visualize transcribed ribosomal RNA genes (2), but these spreading conditions (low-salt, high-pH) are less well suited to studying the structure of nucleosomes and their folding in higher-order structures. Alternatively, chromatin was analysed in thin sections cut through embedded nuclei and in cryosections of rapidly frozen nuclei (for discussion and references see ref. 3). The results showed that chromatin in nuclei forms compact fibres, which are irregularly shaped along the fibre length and the approximate diameters may scatter in a range of 30 to 40 nm. However, since the sections are only 50 to 100 nm thick, and since chromatin fibres are tightly packed in nuclei, it is impossible to follow the same chromatin fibre over long distances. Furthermore, these approaches did not resolve nucleosomes in chromatin fibres, and details on how nucleosomes are connected, arranged, and folded in the compact fibres remain obscure. A serious problem in structural analyis is the insolubility of long chromatin at higher ionic strengths (> 60 mM NaCl) and under physiological conditions. Therefore, protocols were developed to enzymatically cleave and to solubilize chromatin by lysing nuclei in low-salt solutions (4). These fragments are tractable for biochemical and EM analysis.

A protocol is described in this chapter which allows one to visualize soluble chromatin fragments as nucleoprotein complexes in a transmission electron microscope (TEM). This technique is suitable for the examination of nucleosome distribution along the DNA and condensation properties of chromatin, but it is equally well suited for testing chromatin samples after reconstitution (Chapters 8–10) and for characterizing chromatin substrates that are used for *in vitro* transcription, replication, or DNA-repair experiments. This technique has been described previously (5–8). Briefly, chromatin fragments are generated by lysis of the nuclei after nuclease digestion (Chapter 1, section 3) or by reconstitution of DNA with histones (Chapters 8 and 9). Chromatin may be purified on sucrose gradients to remove unbound proteins and reduce background in EM, or to dissociate chromosomal proteins such as histone H1 (Chapter 1). The nucleoprotein complex is fixed with glutaraldehyde in solution and immobilized by adsorption on carbon-coated EM grids. After washing and dehydration, chromatin is contrasted by shadowing with platinum–carbon and EM may be performed in a conventional transmission mode. Since all steps may affect the structural features of the specimen, they need to be carefully controlled. It is suggested that chromatin is kept for as long as possible in solution, which allows biochemical and biophysical controls of the material (e.g. by sedimentation, nuclease digestion, analysis of protein–DNA composition and integrity). From the moment of adsorption onto the support grid, controls can be done by EM only. In these cases, it is recommended that material with known properties is included (e.g. rat liver chromatin, H1-depleted chromatin, DNA) as standards for comparative analysis.

EM of chromatin from rat liver (Chapter 1, *Protocol 1*, and Chapter 5, *Protocol 1*) digested with micrococcal nuclease (MNase) and fractionated in sucrose gradients (see Chapter 1, *Protocols 8* and *9*) is described in this chapter.

2. Protocols

Protocol 1. Preparation of carbon-coated grids and of Alcian-blue treated grids

Equipment and reagents

- High Vacuum Evaporator BA 360 M with electron gun, quartz crystal, thin film monitor QSG 201, and control unit EVM052 (Bal-Tec)
- Mica, carbon, platinum (90%)–carbon (10%) rods, EM support grids (Bal-Tec)
- 0.22 μm filter (Millipore)
- 0.2% Alcian blue 8GX (Serva) in 3% acetic acid stock solution (keep at room temperature)
- Scotch solution: 20–30 cm Scotch tape dissolved in chloroform. (The cellophane support of the tape does not dissolve)
- Wire-mesh support stand[a]

A. *Preparation of carbon-coated grids (C-grids)*

1. Cleave a piece of mica (about 2 cm × 5 cm) with forceps and place it with the freshly cleaved surface facing upwards at a distance of about 12 cm below the carbon evaporator gun[a]. Place the quartz close to the mica. Turn the shutter to cover the mica. Pre-heat the filament of the

electron gun at a vacuum of about 5×10^{-6} torr. When the evaporation rate is constant, open the shutter and deposit 60–70 Hz (measured with the thin film quartz monitor) on the mica. The thickness of the carbon film is about 6–7 nm.

2. Place the mica, with the carbon film facing upwards, on a wet filter paper (soaked in redistilled water) in a covered Petri dish and process within the next hour.
3. Place EM support grids on a filter paper and rinse them with one to two droplets of Scotch solution. The glue of the Scotch tape helps to keep the carbon film attached to the copper grid.
4. Place a filter paper (approximately the same size as the carbon film) on a wire-mesh support stand submerged in redistilled water[a]. Put the copper grids, glue side up, on the filter paper.
5. Carefully lower the carbon-coated mica (carbon side up) into the water at an angle of about 45° until the carbon film is completely released and is floating on the water surface.
6. Move the carbon film onto the grids by carefully sucking off the water, using an aspirator.
7. Remove the filter paper with the carbon-coated grids and place it on a dry filter paper in a glass Petri dish and incubate at 160 °C for 3–4 min. Store the carbon-coated grids (C-grids) at room temperature and use them within 2–15 h. During this time, the C-grids maintain good adsorption properties for double-stranded DNA and chromatin in the appropriate buffers (see below).

B. *Preparation of Alcian-blue coated carbon grids (ABC-grids)*

The adsorption quality of C-grids may vary between different laboratories and may change with age. Coating the C-grids with Alcian blue may be used to improve the adsorption quality of 'bad' C-grids (7). The disadvantage is a rougher background of ABC-grids. This may make it necessary to stain DNA and chromatin with uranyl acetate prior to shadowing .

1. Immediately prior to use, dilute the Alcian-blue stock solution with redistilled water to a final concentration of 2×10^{-3} % Alcian blue. Filter through a 0.22 μm filter.
2. Float C-grids, with the carbon side down, on 2×10^{-3} % Alcian blue for 5 minutes at room temperature.
3. Wash off the excess Alcian blue by floating the grids with the C side down on redistilled water for 10 minutes.
4. Dry the grids (ABC-grids) on a filter paper and use them within the next hour.

[a] See *Figure 1*.

These methods are illustrated in *Figure 1*.

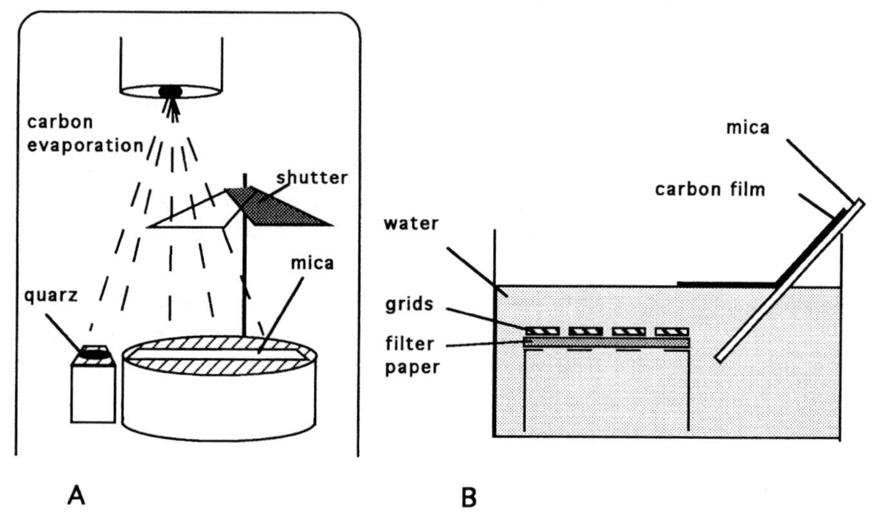

Figure 1. Preparation of carbon-coated grids (C-grids). (A) Evaporation of carbon on a freshly cleaved mica surface. The thickness is monitored by a quartz sensor. (B) The carbon film is released from the mica onto a water surface and lowered down on the copper grids by sucking off the water.

Protocol 2. Preparation of soluble chromatin on C- and ABC-grids

Equipment and reagents

- Dialysis bags (Spectra/Por, boiled in EDTA, rinsed and stored in redistilled water)
- 0.5 M TEACl (triethanolamine; adjust to pH 7 with HCl)
- 0.5 M EDTA, pH 7
- 25% glutaraldehyde (Merck); keep at 4 °C under nitrogen
- BAC stock solution: 0.2% in redistilled water; store at room temperature. (BAC, benzyl-dimethylalkylammonium chloride: *n*-alkyl mixture: $C_{12}H_{26}$, 60%; C_{14},H_{29}, 40%; Bayer, Leverkusen. It is currently unavailable, but can be obtained from our Laboratory.)
- 98% ethanol (Merck)
- Salt-series buffers:
  - 1 mM: 1 mM TEACl, 0.2 mM EDTA
  - 10 mM: 5 mM TEACl, 0.2 mM EDTA, 40 mM NaCl
  - 60 mM: 5 mM TEACl, 0.2 mM EDTA, 60 mM NaCl
  - 100 mM: 5 mM TEACl, 0.2 mM EDTA, 100 mM NaCl.
- Adsorption buffer: 1 mM or 5 mM TEACl, pH 7, 0.2 mM EDTA, 2×10^{-4} % BAC
- Redistilled water
- EM calibration grid (2160 lines/mm; Bal-Tec)

Method

1. Dialyse chromatin samples (about 20–100 μg DNA/ml) extensively against the salt-series buffers at 4°C (2×100 volumes; cover the dialysis vessels with Parafilm to minimize the uptake of carbon dioxide and a drop in pH).

2. Fix the samples by dialysis for about 16 h at 4°C in salt-series buffers containing 0.1% freshly added glutaraldehyde. The fixed chromatin may be stored at 4°C and used for electron microscopy for a few weeks. (Do not freeze the samples.)
3. For EM preparation, dilute the fixed chromatin samples into adsorption buffer and adjust to room temperature for about 30 min. For extended chromatin (e.g. fixed in a low-salt buffer), or chromatin without histone H1, use a final concentration of about 0.5–2 μg DNA/ml. For compact chromatin (e.g. fixed in a high-salt buffer), use a final concentration from 5 to 10 μg DNA/ml.
4. Place C-grids (or ABC-grids) on a sheet of Parafilm, carbon side up, and immobilize them by carefully pressing the edge of the grids to the Parafilm with forceps.
5. Place a 5–20 μl droplet of chromatin on the grid and allow the chromatin to adsorb for 5 min[a].
6. Wash off the droplets and excess salt by floating the grids upside down on redistilled water for 10 min.
7. Dehydrate the samples by dipping the grids into 98% ethanol for about 3 sec.
8. Air-dry the samples on a filter paper.
9. For shadowing, place the grids on a rotary stage about 12 cm from a platinum–carbon evaporation gun. Adjust the angle between the beam centre and the carbon film to 7°. Perform rotary shadowing at a constant rate and at a pressure below 10^{-4} torr. Measure the evaporation with the quartz device placed perpendicular to the beam[a]. 500 Hz and 750 Hz are optimal for C-grids and ABC-grids, respectively.
10. Perform transmission electron microscopy at 75–100 kV and at a magnification of about 20 000 (checked with a calibration grid). The resolution of the shadowed specimens is about 2–3 nm as defined by the distance between platinum clusters.

[a] See *Figure 2*. Too brief adsorption (e.g. by touching a droplet of chromatin with a carbon grid) may result in partially attached chromatin filaments, and artificial stretching or compaction might occur during subsequent steps. Longer adsorption times are not critical

These methods are illustrated in *Figure 2*.

3. Discussion

3.1 Variability of chromatin appearance

The protocol described here was optimized to visualize soluble chromatin from mononucleosomes to >100 nucleosomes, but it may also be used to

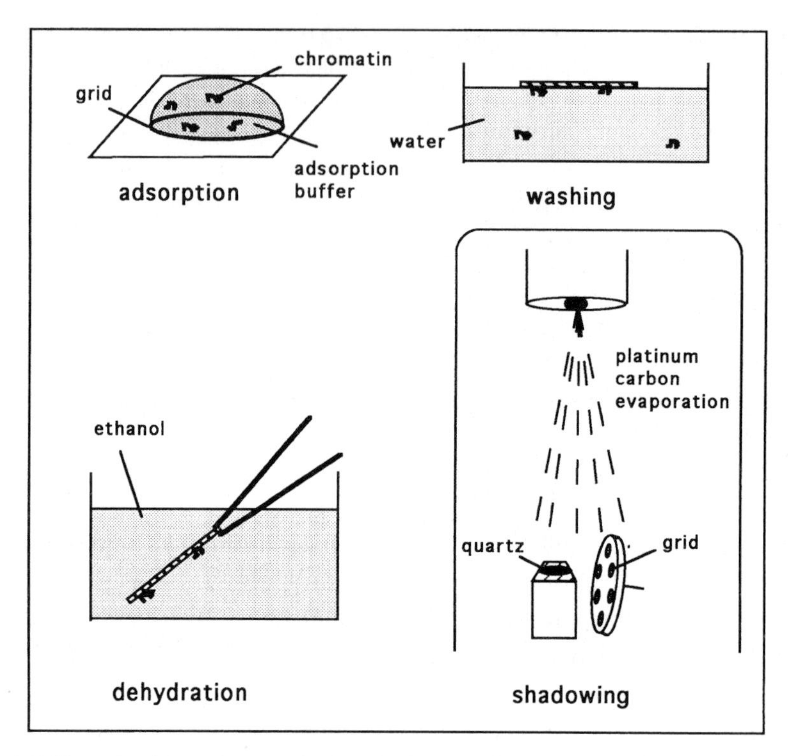

Figure 2. Steps in chromatin preparation. Fixed chromatin in adsorption buffer is placed on an EM grid (adsorption) for 5 minutes, the grid is washed by floating it on water (washing) for 10 minutes. The samples are then dehydrated in ethanol (dehydration) for a few seconds, air-dried on filter paper, and contrasted by rotary shadowing at a fixed angle (shadowing).

investigate other protein–DNA complexes, e.g. the interactions of DNA-gyrase with DNA (12). Chromatin is very sensitive to changes in ionic strength, ion composition, protein/DNA composition, and pH. This is illustrated in *Figures 3* and *4*.

In '1 mM' salt-series buffer, pH 7, rat liver chromatin containing all histone proteins and non-histone chromosomal proteins appears as a loose filament with a width of about one to two nucleosomes (*Figure 3a*). Nucleosomes are frequently arranged in a 'zigzag', since the linker DNA enters and leaves nucleosomes on the same side (*Figure 4e*) and since nucleosomes may adsorb with their flat surfaces on the carbon grids. This chain of nucleosomes is called a 'nucleosome filament' (6). The nucleosome filaments are distinct from 'nucleofilaments', which are continous 10 nm filaments obtained by negative-staining with uranyl acetate (13).The compaction of DNA in nucleosome filaments is about six- to sevenfold. (This is calculated as the length of protein-free DNA/length of packaged DNA: one nucleosome contains about 200 bp DNA with a length of about 68 nm (= 200 bp $\times$ 0.34 nm/bp). In nucleosome

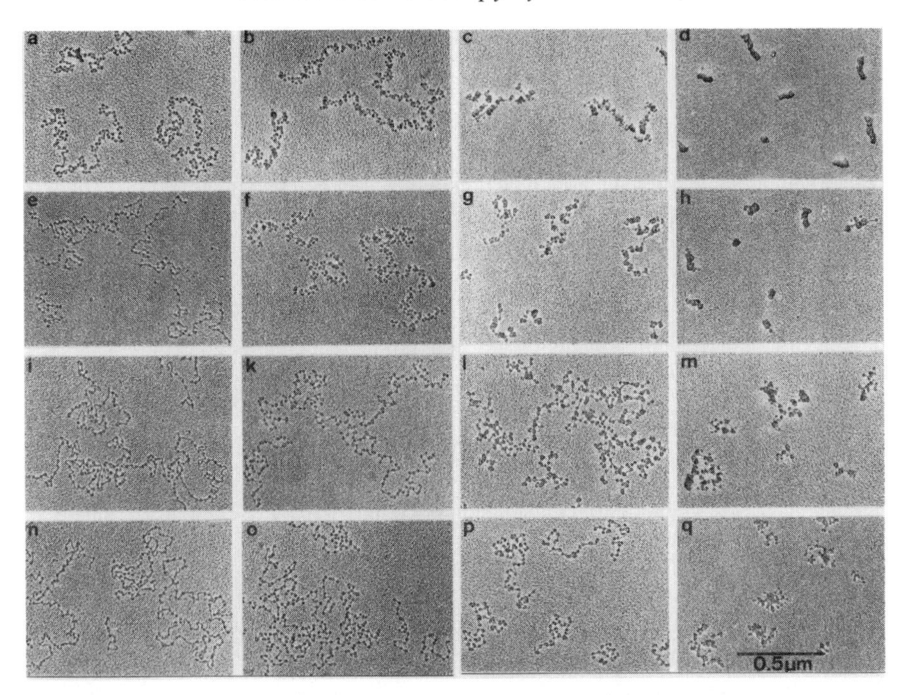

Figure 3. Variability of chromatin structures as a function of salt concentration and protein/DNA composition. Soluble rat liver chromatin fractionated in '10 mM NaCl' (a–d). Chromatin after removal of non-histone chromosomal proteins (NHCP) by sedimentation in 300 mM NaCl (e–h). Chromatin after removal of NHCP and a fraction of H1 by sedimentation in 350 mM NaCl (i–m). Chromatin depleted of histone H1 and NHCP by sedimentation in 500 mM NaCl (n–q). Chromatin was fixed with 0.1% glutaraldehyde in '1 mM' (a, e, i, n), '10 mM' (b, f, k, o), '40 mM' (c, g, l, p), '100 mM' (d, h, m, q). (Adapted from Thoma and Koller (10), with permission.)

filaments, there is about one nucleosomes per 10 nm filament length. The packaging ratio is 68 nm/10 nm).

In '10 mM', '40 mM', and '100 mM' salt-series buffers, chromatin condenses into more compact fibre-like structures ('salt-dependent condensation'; *Figures 3b–d*). After removal of non-histone proteins in 300 mM NaCl, the nucleosome filament appears to be more extended in '1 mM', but closes into fibre-like structures at higher ionic strength (*Figures 3e–h*).

Additional removal of H1 leads to a destabilization and partial unfolding of nucleosomes (loss of beads) in '1 mM' (*Figures 3n* and *4c*). At '10 mM', nucleosomes appear as 'beads-on-a-string' (*Figures 3o* and *4d*). When compared with the zigzag, 'beads-on-a-string' are consistent with a partial unwrapping of the ends of nucleosomal DNA. In this case, the mass/unit length corresponds to about one nucleosome per 20–30 nm, which corresponds to a packing ratio of approximately 2–3. H1-depleted chromatin does condense at higher ionic strength, but it does not fold into distinct fibre-like structures (*Figures 3p,q*).

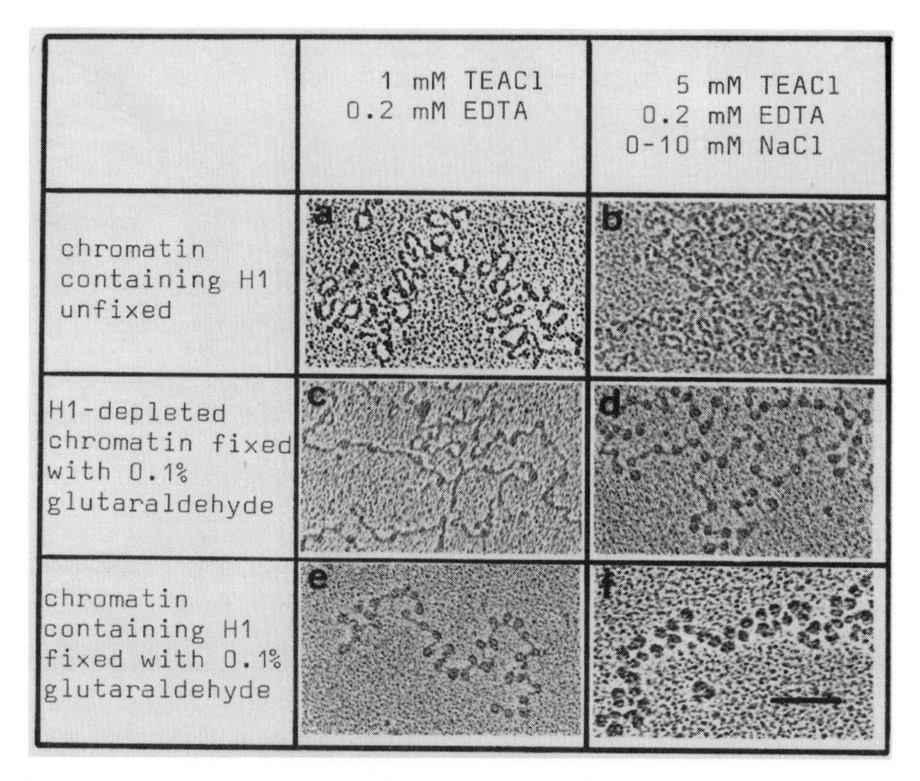

Figure 4. Chromatin disruption and unfolding. Nucleosomes are disrupted when unfixed chromatin is adsorbed on C-grids (a, b). H1-depleted chromatin fixed in low-salt solutions shows (partially) unfolded nucleosomes (c). H1-depleted chromatin fixed in '10 mM' shows 'beads-on-a-string' (d). Chromatin fixed in '1 mM' frequently shows nucleosomes arranged in a 'zigzag', since DNA enters and leaves nucleosomes on the same side (e). Chromatin fixed in '10 mM' shows nucleosomes in contact (closed 'zigzag'). Bar is 0.5 μm. (Adapted from Thoma *et al.* (6), with permission.)

Loss of fibre formation is already observed when histone H1 is partially removed (e.g. at 350 mM, *Figure 3m*).

Alkaline pH (pH 9–10) is used in chromatin preparations according to Miller and Beatty (1,2). High pH (pH 9 or greater), however, promotes expansion of the nucleosome filament and partial disruption of nucleosomes to give a 'beads-on-a-string' type appearance, pronounced unfolding of nucleosomes in H1-depleted chromatin at low ionic strength, and disruption of higher-order chromatin fibres (7).

3.2 Specimen preparation

A general criterion of quality control requires that a biochemically homogeneous population of chromatin results in a homogeneous population on the

EM grid (*Figure 3*). Several parameters of specimen preparation which may affect the appearance of chromatin need to be controlled (for more details see ref. 6).

3.2.1 Fixation

During adsorption, washing, staining, and dehydration, forces may act on chromatin structures and generate structural changes and loss or rearrangement of components. Indeed, when unfixed chromatin is adsorbed on C-grids according to the procedure described above, irregular thickness DNA-like filaments are observed consistent with unravelled nucleosomes or partially dissociated nucleosomes (*Figures 4a, b*). This problem is solved by the chemical fixation of chromatin in solution using glutaraldehyde, which preferentially crosslinks proteins. For interpretation, it is important to realize that fixation can shift the equilibrium between folded and unfolded states towards the more compact, folded state. Chromatin fixed in an extended state, e.g. in low ionic strength, does still condense when adjusted to higher ionic strength, but chromatin fixed in a compact state is less likely to open up in lower ionic strength. This allows one to use one common low-salt buffer for adsorption.

3.2.2 Adsorption, washing

Firm adsorption of chromatin to the support film is required to prevent structural changes during the subsequent steps. The following criteria are used to judge adsorption: (a) a biochemically homogeneous chromatin sample should result in a homogeneous population of chromatin structures on the EM grid. The grids should be covered with equal density (no patches). (b) Stretched filaments indicate insufficient adhesion and may not be used for structural interpretations. (c) When grids with extended chromatin are washed in 1 mM Mg^{2+} and dried, no compaction should occur. On our C-grids, chromatin and DNA adsorb well in 1–5 mM TEACl, pH 7, 0.2 mM EDTA containing 10 mM (or more) NaCl. To visualize chromatin at a lower ionic strength, a minimum of 2×10^{-4} % BAC is required to provide sufficient adhesion. Since the quality of carbon films is variable in different hands and laboratories, procedures were developed to coat the surface with positively charged molecules such as amylamine (14), polylysine (15), or Alcian blue (7). However, these molecules do interact with chromatin and DNA, and it is not known whether, and to what extent, these molecules may alter the local composition and structure of chromatin. Indeed they may have a stabilizing effect, since unfixed chromatin adsorbed to ABC-grids or amylamine-coated grids does not disintegrate to the same extent as on C-grids. A disadvantage of coated grids are a rougher background which may require positive staining of DNA prior to metal shadowing. Furthermore, patchy coating of the grids may lead to a different appearance of chromatin in different areas of a grid.

3.2.3 Dehydration

Ethanol dehydration of DNA and chromatin may induce folded forms with diameters of ~ 30 nm, which in the case of chromatin could easily be mistaken for 30 nm thick fibres. To control for dehydration artefacts, specimens may be freeze-dried by rapidly freezing in liquid nitrogen and slowly drying *in vacuo* on a cold stage. In our samples, no significant differences were observed between ethanol-dehydrated/air-dried and freeze-dried specimens (6).

3.2.4 Staining and metal shadowing

For conventional, transmission electron microscopy of biological specimens contrast enhancement by staining or shadowing with heavy metals is necessary.

i. Positive staining with uranyl acetate

After adsorption on C- or ABC-grids, the grids are floated for 1–2 minutes on a freshly prepared and filtered 1% uranyl acetate solution (in water). Uranyldioxide binding to DNA results in positive staining of chromatin. Results should be interpreted with caution. Since chromatin is very sensitive to divalent cations, binding of uranyldioxide might alter the local structure. Moreover, selective binding to DNA emphasizes the DNA component of chromatin.

ii. Negative staining with uranyl acetate

After adsorption of chromatin, the grids are rinsed with three droplets of uranyl acetate (1% uranyl acetate in water, freshly filtered), the excess solution is removed with filter paper, and the grids are air-dried. Negative staining depends on the surface properties of the grids. Usually, C- and ABC-grids result in positive staining of adsorbed chromatin and DNA. However, if the grids are overloaded and chromatin is not firmly attached, the chromatin will be embedded in the uranyl and appear as 10 nm wide continous filaments called 'nucleofilaments' in which nucleosomes are apparently packed face-to-face and stabilized by the uranyl environment (13).

iii. Platinum–carbon shadowing

Heavy metals (in our case platinum) are deposited on the surface of the specimen and support film *in vacuo*. While positive and negative staining occur in solvents and may result in structural changes of the chromatin due to chemical interactions, metal deposition will not alter the structures. However, metal deposition is not random. Preferential nucleation sites on the substrate and support film may lead to the formation of platinum kernels and, hence, to 'decoration' of a structure. Moreover, too much metal deposition may mask structural details. After rotary shadowing under the above conditions, nucleosomes with a diameter of 11 nm appear as large as 16 nm. The height of the specimen may be measured by metal shadowing from a fixed angle. In this case, latex beads of known diameters (90 nm; Agar Scientific Ltd, Stansted

Essex, UK) are co-prepared by floating the grids, after the washing step, on a suspension of 10^{12} latex beads/ml in water.

iv. Mixing experiments

In view of the numerous parameters that affect the appearance of chromatin in the electron microscope, only some of which are inherent properties of chromatin, while others may be related to specimen preparation, it is therefore suggested that co-preparation or 'mixing experiments' be performed (*Figure 5*) (6). Since chromatin samples are fixed in solution, they can be mixed together in the same adsorption buffer and co-prepared on the same grid. This procedure keeps the parameters during specimen preparation for both samples constant. For example, when chromatin fixed in 'high-salt' is mixed with chromatin fixed in 'low-salt' conditions (*Figure 5A*), compact fibres and low-salt filaments are observed in the appropriate ratio. When chromatin containing H1 fixed in 'low-salt' is mixed with H1-depleted chromatin fixed in 'low-salt' (*Figure 4b*), the characteristic nucleosome filaments and partially unfolded filaments are observed (*Figure 4b*).

4. Alternative techniques

4.1 Chromatin analysis by psoralen crosslinking

Psoralen crosslinking is used as an indirect approach to display the arrangement of nucleosomes along the DNA (for a detailed protocol see ref. 8). Psoralen derivatives intercalate in the DNA helix and form covalent crosslinks between pyrimidine bases of opposite strands when irradiated with ultraviolet light (366 nm). For chromatin analysis, whole cells, purified chromatin, or reconstituted chromatin samples are treated with psoralen and 366 nm UV light. Psoralen crosslinks are preferentially formed in linker DNA and in nucleosome-free DNA regions, such as the Simian Virus 40 origin of replication or transcribed ribosomal RNA genes. Nucleosomes are resistant to crosslinking. The DNA is purified, denatured, adsorbed on EM support grids using a protein-free sprading technique, stained with uranyl acetate, shadowed with platinum–carbon, and visualized by TEM. Crosslinked DNA cannot denature and appears as double-stranded DNA, while DNA resistant to crosslinking is denatured and appears as single-stranded DNA. Hence, a nucleosomal region appears as an alternating pattern of double-stranded DNA (linker DNA) and single-stranded bubbles (nucleosomal DNA). Transcribed rRNA genes, which are devoid of nucleosomes, appear as long, double-stranded regions. While this approach does not allow one to obtain structural information above the nucleosome level, it does allow one to measure and quantitate nucleosomal bubbles, linker DNA, and transcribed regions. This method has been successfully applied to characterize chromatin structures of replicating minichromosomes, ribosomal RNA genes, and *in vitro* replicated chromatin (16–18). Furthermore, it is possible to simultaneously crosslink nascent RNA

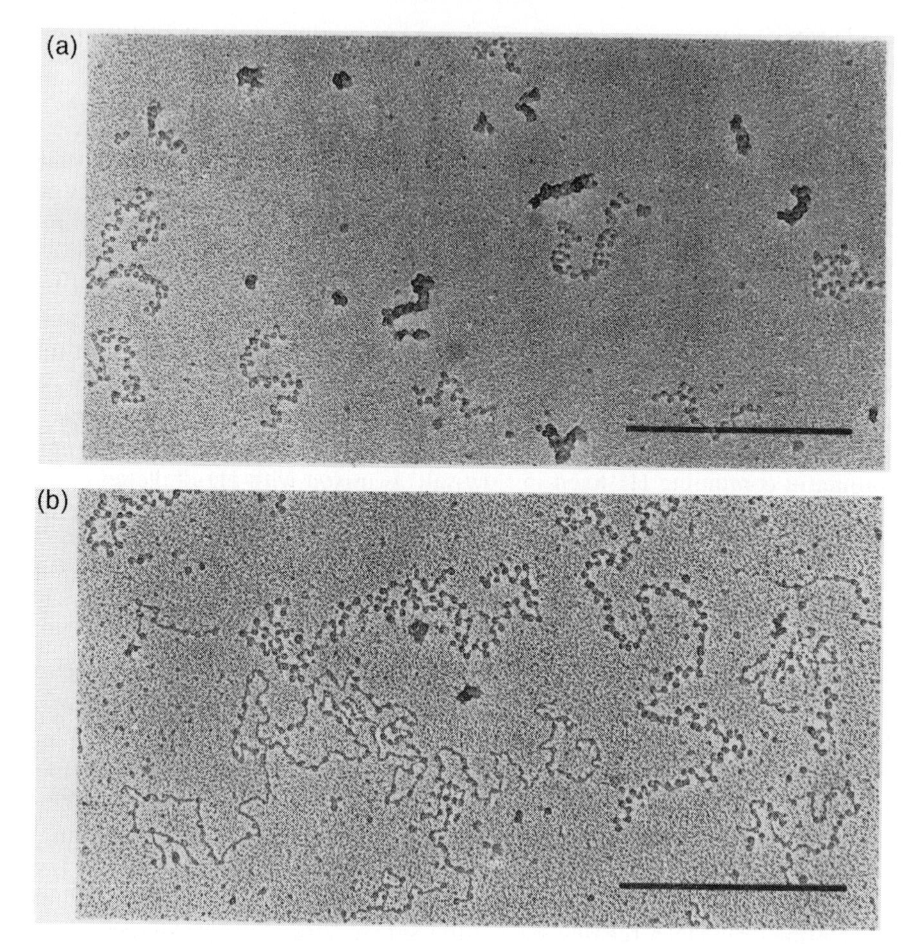

Figure 5. Mixing experiments. (a) Chromatin was fixed with 0.1% glutaraldehyde in '1 mM' and '100 mM'salt-series buffers, respectively, diluted in the same adsorption buffer, and co-prepared for EM. The characteristic compact fibres and extended nucleosome filaments are observed. (b) Chromatin and H1-depleted chromatin was fixed in '1 mM' salt buffer, mixed in the same adsorption buffer, and co-prepared for EM. The characteristic nucleosome filaments and the unfolded nucleosomes are observed. (Adapted from Thoma *et al.* (6), with permission.)

for analysis of transcribed chromatin (19). In addition, since the DNA is not destroyed and since psoralen crosslinks are reversible when DNA is irradiated at 254 nm, the DNA remains accessible for molecular biological analyses (20).

4.2 Cryoelectron microscopy

In order to avoid adsorption, staining, and dehydration, and to visualize chromatin as it occurs in solution, chromatin samples are rapidly frozen in thin layers of solvent. The specimen is included in vitrified ice and transferred into the electron microscope equipped with a cold stage (21). So far, this technique may not be used for routine analyses of chromatin samples, since it requires special equipment. Furthermore, it is difficult to analyse and interpret the data due to the low contrast of unstained biological molecules. Furthermore, the thickness of vitrified specimens is about 50 nm. Hence, longer chromatin fragments will be oriented during vitrification.

4.3 Scanning force microscopy (SMF)

Recently, SFM has been applied for visualizing chromatin. SFM uses a sharp tip mounted on a flexible cantilever to touch and create images of a surface (for a review see ref. 22). Compared with the conventional EM preparation procedure described above, SFM has the potential to analyse unstained samples in a hydrated state, and no transfer *in vacuo* is required. Limitations of SFM of biological specimens arise from the direct interaction between the tip and sample.

As a preparative limitation, the specimens need to be firmly immobilized on a surface. The adhesion to the surface must be stronger than the forces applied by touching the sample. Otherwise, the specimen moves and structures change when touched by the tip and images blur out. This makes it basically impossible to visualize a structure in solution. It will also make it difficult to obtain structural information on large, flexible objects such as chromatin fibres. Since only that part of the fibre which is adsorbed on the surface is immobilized, the other side may still be mobile unless chromatin is firmly fixed by chemical means. So far, humidities below 30% are required to obtain stable imaging conditions for protein/DNA complexes (22).

A technical limitation comes from the finite dimension, size, and shape of the tip. Compared with nucleosome dimensions (11 nm $\times$ 0.6 nm) or DNA diameters (2 nm), even the smallest tips with radii of curvature of 8–12 nm are still very large. This leads to an overestimation of lateral dimensions (22) and, with respect to chromatin, prevents the tip from entering between the grooves of closely spaced nucleosomes as they occur in chromatin fibres.

For the preparation of chromatin samples, chromatin is adjusted to 'low-salt' (same buffers as described above for TEM), adsorbed on a glass or mica surface, which in the case of mica is pretreated with spermidine, rinsed with water, and dried with nitrogen gas (23, 24). The results obtained with chromatin fragments at low ionic strength and on H1-depleted chromatin, which both show expanded chromatin structures, are promising (23, 24). They have confirmed the previous observations of low-salt structures by transmission electron microscopy (5, 6). Second, the SFM results further support the conclusion that dehydration in ethanol followed by analysis *in vacuo* is not a severe limitation for TEM.

Acknowledgement

This article was supported by a grant of the Schweizerischer Nationalfonds zur Förderung der wissenschaftlichen Forschung.

References

1. Olins, A. L. and Olins, D. E. (1974) *Science*, **183**, 330.
2. Miller, O. L. and Beatty, B. R. (1969) *Science*, **164**, 955.
3. Woodcock, C. L. (1994) *J. Cell Biol.*, **125**, 11.
4. Noll, M. and Kornberg, R. D. (1977) *J. Mol. Biol.*, **109**, 393.
5. Thoma, F. and Koller, T. (1977) *Cell*, **12**, 101..
6. Thoma, F., Koller, T., and Klug, A. (1979) *J. Cell Biol.*, **83**, 403.
7. Labhart, P., Thoma, F., and Koller, T. (1981) *Eur. J. Cell Biol.*, **25**, 19.
8. Sogo, J. M. and Thoma, F. (1988) In *Methods in enzymology* (ed. R. Kornberg and P. Wassarman), Vol. 170, p. 142. Academic Press, London.
9. Hewish, D. R. and Burgoyne, L. A. (1973) *Biochem. Biophys. Res. Commun.*, **52**, 504.
10. Thoma, F. and Koller, T. (1981) *J. Mol. Biol.*, **148**, 709.
11. Noll, H. and Noll, M. (1989) In *Methods in enzymology* (ed. R. Kornberg and P. Wassarman), Vol. 170, p. 55. Academic Press, London.
12. Rau, D. C., Gellert, M., Thoma, F., and Maxwell, A. (1987) *J. Mol. Biol.*, **193**, 555.
13. Finch, J. T. and Klug, A. (1976) *Proc. Natl Acad. Sci. USA*, **73**, 1897.
14. Dubochet, J., Ducommun, M., Zollinger, M., and Kellenberger, E. (1971) *J. Ultrastruct. Res.*, **35**, 147.
15. Williams, R. C. (1977) *Proc. Natl Acad. Sci. USA*, **74**, 2311.
16. Sogo, J. M., Ness, P. J., Widmer, R. M., Parish, R. W., and Koller, T. (1984) *J. Mol. Biol.*, **178**, 897.
17. Sogo, J. M., Stahl, H., Koller, T., and Knippers, R. (1986) *J. Mol. Biol.*, **189**, 189.
18. Gruss, C., Wu, J. R., Koller, T., and Sogo, J. M. (1993) *EMBO J.*, **12**, 4533.
19. De Bernardin, W., Koller, T., and Sogo, J. M. (1986) *J. Mol. Biol.*, **191**, 469.
20. Conconi, A., Widmer, R. M., Koller, T., and Sogo, J. M. (1989) *Cell*, **57**, 753.
21. Dubochet, J., Adrian, M., Schultz, P., and Oudet, P. (1986) *EMBO J.*, **5**, 519.
22. Bustamante, C., Keller, D., and Yang, G. (1993) *Curr. Opin. Struct. Biol.*, **3**, 363.
23. Zlatanova, J., Leuba, S. H., Yang, G. L., Bustamante, C., and Van Holde, K. (1994) *Proc. Natl Acad. Sci. USA*, **91**, 5277.
24. Leuba, S. H., Yang, G., Robert, C., Samori, B., Van Holde, K., Zlatanova, J., and Bustamante, C. (1994) *Proc. Natl Acad. Sci. USA*, **91**, 11621.

7

Counting nucleosome cores on circular DNA using topoisomerase I

DAVID J. CLARK

1. Introduction

The nucleosome is the basic structural repeat of chromatin. It contains a nucleosome core, a single molecule of histone H1, and a segment of linker DNA. The nucleosome core is composed of a central octamer of core histones (two molecules each of the four core histones: H2A, H2B, H3, and H4) and about 150 bp of DNA which is tightly wrapped around the octamer, forming nearly two complete turns. A nucleosome core protects one negative DNA supercoil from relaxation by topoisomerase I (1). This was shown by counting the average number of nucleosome cores reconstituted on a plasmid *in vitro* using the electron microscope and then comparing this with the average number of supercoils protected from relaxation in the same samples. This fact is now used routinely to determine the average number of nucleosome cores formed on a circular DNA: the DNA between nucleosome cores is relaxed using topoisomerase I, plasmid DNA is extracted, and the number of negative supercoils is measured in a gel containing chloroquine. The number of nucleosome cores is equal to the number of negative supercoils protected from relaxation.

In this chapter, this method for counting nucleosome cores *in vitro* is described in some detail. It is a sensitive and accurate assay for measuring the number of nucleosome cores formed on circular DNA. (It cannot be used to count nucleosome cores on linear DNA, because this is not topologically constrained.) It has also been used to look for conformational changes in the nucleosome due to core histone acetylation (2, 3) or positive supercoiling (4, 5), or to measure DNA supercoiling in subnucleosomal particles such as the H3/H4 tetramer (6). It is also possible to count nucleosomes in circular minichromosomes isolated from natural sources, e.g. SV40 minichromosomes (7, 8) or artificial yeast minichromosomes (9).

1.1 DNA supercoiling

The change in linking number (ΔLk) (i.e. the change in the number of times the two strands of the DNA double helix are intertwined) for a circular DNA is given by:

$$\Delta Lk = \Delta Wr + \Delta Tw, \qquad [1]$$

where ΔWr is the change in DNA writhe (which describes the path of the DNA duplex in space) and ΔTw is the change in DNA twist (the number of base pairs per turn of the DNA helix, which equals 10.5 bp/turn for B-DNA in solution) (10). The helical twist of DNA depends on temperature (11) and on salt concentration.

The superhelix density, σ, is defined as the number of supercoils per helical turn of DNA:

$$\sigma = \Delta Lk/(\text{no. of helical turns in the DNA}) = \Delta Lk/(PL/10.5) \qquad [2]$$

where PL is the length of the plasmid in base pairs. This is a useful measure because it is independent of DNA length and therefore the degree of supercoiling of circular DNAs of different sizes can be compared (see below).

1.2 DNA supercoiling in the nucleosome core

The ΔLk measured for nucleosome core formation is -1.0 (1). However, nearly two negative supercoils of DNA are present in the crystal structure of the nucleosome core (12), suggesting a change in writhe closer to -2 than -1; this is the basis of the so-called linking number paradox. Measurements of DNA twist in the nucleosome core showed that some of the DNA on the nucleosome surface is overwound to 10.0 bp/turn, contributing a *positive* change in twist which partially compensates for the negative writhe, according to *Equation 1*, and goes some way towards resolving the paradox (see refs 12 and 13 for discussion). Although the interpretation of the observed value of $\Delta Lk = -1$ in terms of the twist and writhe of nucleosomal DNA is important in experiments designed to detect conformational changes in nucleosome cores, a full understanding of the structural basis of the linking number change is unnecessary for simple measurement of the number of nucleosome cores. The value of $\Delta Lk = -1$ per nucleosome core is generally accepted.

2. The relaxation reaction

To determine the number of nucleosome cores in chromatin reconstituted *in vitro* using purified core histones and DNA, it is necessary to incubate it with topoisomerase I. This can either be obtained commercially (Gibco/BRL) or it can be prepared in a crude form relatively cheaply and easily from chicken erythrocyte nuclei (6). One unit of topoisomerase activity is defined as the amount of extract required to relax 1 μg of pBR322 in 30 min at 37 °C in 0.2 M NaCl, 20 mM Tris–HCl, pH 8, 1 mM Na-EDTA. Similar definitions of activity are given by the manufacturers of commercial preparations of topoisomerase I. If the chromatin is assembled in crude extracts from *Xenopus* eggs or

Drosophila embryos (e.g. refs 15, 16), endogenous topoisomerase activity is usually sufficient to relax the DNA during nucleosome assembly.

2.1 Relaxation of reconstituted chromatin with topoisomerase

A protocol for the relaxation of reconstituted chromatin samples using nicking–closing extract (or commercial topoisomerase I) is given below. Chromatin is relaxed at a slower rate than nucleosome-free DNA under the same conditions and so the time of incubation is extended to ensure complete relaxation. However, at very high nucleosome densities (approaching one nucleosome per 150 bp when nucleosomes are closely packed), it is difficult to relax the DNA to completion, presumably because the DNA is inaccessible to topoisomerase. However, this problem seldom arises because such a nucleosome density is greater than physiological values (one per 165–240 bp, depending on the tissue source).

Protocol 1. Relaxation of reconstituted chromatin with topoisomerase I

Equipment and reagents

- Topoisomerase I or nicking–closing extract
- Speed-Vac concentrator (Savant)
- Buffer 1: 0.2 M NaCl, 20 mM Tris–HCl, pH 8.0, 1 mM Na-EDTA[a]
- 10% (w/v) sodium dodecyl sulfate (SDS)
- 3 M sodium acetate (pH adjusted to 7.0 with acetic acid)
- Phenol/chloroform (1:1)
- Ethanol and 70% (v/v) aqueous ethanol
- Microcentrifuge
- TPE sample buffer: 40 mM Tris-base, 30 mM NaH_2PO_4, 1 mM Na-EDTA, pH 8.2, 10% (v/v) glycerol
- Water bath at 37 °C

Method

1. Dilute 0.8 μg reconstituted chromatin to 80 μl in buffer 1[a]. This corresponds to a DNA concentration of 10 μg/ml, but concentrations from 5–50 μg/ml are satisfactory.
2. Prepare a control reaction using free plasmid in the same buffer.
3. Add 1 unit nicking–closing extract or topoisomerase I per μg DNA.
4. Incubate for 1 h at 37 °C.
5. Add SDS to a final concentration of 0.2% to stop the reaction and to remove histones from DNA.
6. Add one-tenth volume 3 M sodium acetate (pH 7.0).
7. Extract the aqueous phase with an equal volume of phenol/chloroform (1:1).
8. Extract with an equal volume of chloroform.
9. Precipitate DNA from the aqueous phase by adding 3.5 vols of ethanol.

Protocol 1. *Continued*

10. Leave for 30 min on dry ice.
11. Spin down the precipitated DNA at maximum speed in a microcentrifuge (16 000 *g*) for 15 min.
12. Wash the precipitated DNA once with 500 μl cold 70% (v/v) ethanol: 10 min at maximum speed in the microcentrifuge.
13. Dry the pellet: no more than 5 min in a Speed-Vac concentrator.
14. Dissolve the precipitated DNA in 15 μl TPE sample buffer.

[a] Topoisomerase activity in nicking–closing extract is inhibited at salt concentrations lower than 50 mM. Be careful not to expose the chromatin to transient high concentrations of salt which could cause some histone dissociation.

3. Determination of ΔLk in agarose gels containing chloroquine

ΔLk can be determined by one-dimensional or two-dimensional gel electrophoresis in the presence of chloroquine. The two-dimensional gel system has the advantage that more topoisomers can be resolved and displayed in one gel, facilitating comparison with a test sample; the disadvantages are that only a few samples can be analysed in one gel, the gels take much longer to run, and quantitation is not possible. For these reasons, we routinely employ one-dimensional gels, in which many samples can be compared accurately, side by side, and quantitation by scanning densitometry is possible. In order to measure ΔLk, the correct chloroquine concentration must be chosen to maximize resolution, and a set of standards must be prepared for each plasmid of interest. These procedures are described below.

3.1 Mobility of topoisomers in gels containing chloroquine

Like ethidium, chloroquine is an intercalating agent that unwinds the DNA helix (negative ΔTw). When chloroquine binds to a closed circular DNA molecule, there must be a compensatory *positive* change in ΔWr (*Equation 1*), because ΔLk cannot change if the duplex remains intact. Thus, the binding of chloroquine to DNA reduces its twist and increases its writhe.

A moderately negatively supercoiled plasmid (negative ΔWr) runs as a single fast migrating band in a gel without chloroquine. In gels containing chloroquine at low concentrations, the plasmid binds small amounts of chloroquine, resulting in the loss of some negative supercoils and slower migration in the gel, with some or all topoisomers resolved. At higher chloroquine concentrations, sufficient chloroquine is bound to unwind all the negative supercoils, and the topoisomers are relaxed and migrate with nicked circles ($\Delta Wr = 0$). More chloroquine in the gel results in positively supercoiled topoisomers (positive ΔWr) which are once again resolved as they migrate faster. At still higher chloroquine concentrations they migrate as a single band of high mobility once more. Thus, the mobility of DNA topoisomers in agarose gels is determined primarily by ΔWr: high values of ΔWr, positive or negative, result in a single band of high mobility. Therefore, to resolve topoisomers in a sample of interest, a chloroquine concentration must be chosen which adjusts ΔWr of the sample to a useful value. For most samples, there are two useful ranges of chloroquine concentrations: a lower one at which the topoisomers are resolved as negative supercoils, and a higher one at which they are resolved as positive supercoils. Whilst this can be useful, it also leads to an ambiguity: is the sample migrating as positive or negative supercoils? This can be resolved by running the sample in two gels containing chloroquine at different concentrations; the sign can be deduced from the behaviour of the sample using the argument given above. Alternatively, it is sometimes possible to deduce the sign by careful comparison of the sample with the standard; supercoils of opposite sign migrate slightly differently (the topoisomer bands are not quite aligned).

These points are now illustrated by a discussion of *Figure 1*, in which a set of topoisomer standards was analysed in gels containing chloroquine at 20 μg/ml (*Figure 1A*) and 5 μg/ml (*Figure 1B*):

(a) All the standards are more positively supercoiled at 20 μg/ml (*Figure 1A*) than at 5 μg/ml (*Figure 1B*).
(b) Relaxed DNA (standard 1) would migrate near nicked circles in the absence of chloroquine (not shown), but the topoisomers are nicely resolved at 5 μg chloroquine/ml. However, at 20 μg/ml they bind more chloroquine and run as a single band of high mobility. In both cases the topoisomers are running as positive supercoils (ΔWr is positive).
(c) Standards 2 and 3 ($\Delta Lk = -5$ and -9, respectively) migrate more slowly than standard 1 because they were initially negatively supercoiled. More chloroquine (20 μg/ml; *Figure 1A*) is required to resolve these topoisomers (as positive supercoils).
(d) Standard 4 ($\Delta Lk = -14$) is nicely resolved at 5 μg/ml (*Figure 1B*). Its topoisomer distribution overlaps with that of standard 1 ($\Delta Lk = 0$), but it is migrating as negative supercoils (not enough chloroquine is bound to cancel all 14 of the negative supercoils initially present). That ΔLk is indeed -14 (not 0) can be deduced by comparing the behaviours of these two standards in the two gels (*Figure 1*).
(e) Standard 3 ($\Delta Lk = -9$) migrates with nicked circles at 5 μg/ml and standard 4 ($\Delta Lk = -14$) migrates with nicked circles at 20 μg/ml. Thus, 5 μg/ml chloroquine is sufficient to cancel about nine negative supercoils in a plasmid of this size and 20 μg/ml cancels about 14.
(f) Resolution of topoisomers in standard 8 ($\Delta Lk = -30$) requires a gel containing more chloroquine.

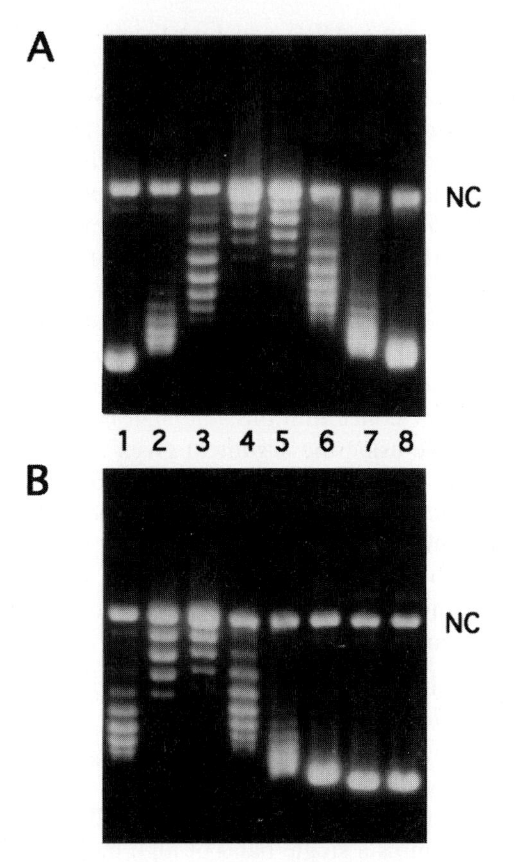

Figure 1. Analysis of a set of topoisomer standards in chloroquine gels. A set of eight topoisomer standards for plasmid pXP10 (3.25 kb) was prepared as described in *Protocol 2*, and analysed in gels containing chloroquine at different concentrations: (A) 20 μg/ml; (B) 5 μg/ml. NC, nicked circle. The average linking numbers (ΔLk) of these standards were determined in a series of gels at different chloroquine concentrations, including the two gels shown here, and are as follows: standard 1, $\Delta Lk = 0$ (by definition; see section 3.2.2). Standard 2, $\Delta Lk = -5$. Standard 3, $\Delta Lk = -9$. Standard 4, $\Delta Lk = -14$. Standard 5, $\Delta Lk = -18$. Standard 6, $\Delta Lk = -22$. Standard 7, $\Delta Lk = -26$. Standard 8, $\Delta Lk = -30$. (See section 3.1 for further discussion.)

3.2 Preparation of topoisomer standards

To determine the linking number of a circular DNA, it is necessary to prepare a series of standards of defined linking number with which to compare the sample of interest. Unfortunately, the standards have to be prepared from the same DNA as the sample DNA (because the DNA size has to be the same for comparison in gels) which entails making a set of standards for each plasmid of interest. An example of a set of such standards is shown in *Figure 1*.

Standards can be prepared using the method of Keller (7) by incubating the plasmid with ethidium bromide at different concentrations and relaxing the

DNA with added topoisomerase I. As explained above, ethidium binding to DNA results in unwinding of the duplex (negative ΔTw) which is compensated for by an equal and opposite change in DNA writhe (positive ΔWr). Topoisomerase I relaxes these positive supercoils, but the DNA remains underwound because ethidium is still bound. The topoisomerase reaction is stopped and the ethidium is extracted. The DNA then restores its twist to the normal value of 10.5 bp/turn and this positive ΔTw is compensated for by a negative change in the writhe, ΔWr, i.e. the plasmid becomes negatively supercoiled. The final number of negative supercoils in the plasmid depends on the amount of ethidium originally bound and therefore on the ethidium concentration.

3.2.1 Range of standards required

The range of standards required depends on the superhelix density expected in the samples of interest; the standards should range from $\Delta Lk = 0$ (relaxed) up to the maximum expected ΔLk. If nucleosomes are packed as closely as possible, each occupying 150 bp of DNA, then the maximum superhelix density possible is $-1/(150/10.5) = -0.07$. This defines the upper limit of the range of standards required. For the 3.25 kb plasmid used in our experiments, it is not possible to form more than 21 nucleosome cores (3250/150). Therefore, standards with $\Delta Lk = 0, -4, -8, -12, -16$, and -20 were prepared. To determine the average linking number of each standard, it is necessary to compare each standard with the next highest in gels containing increased amounts of chloroquine (*Table 1* may be used to estimate the chloroquine

Table 1. Estimation of chloroquine concentration required for optimal resolution of topoisomers

Expected nucleosome density (bp/nucleosome[a])	Expected superhelix density[b] (σ supercoils per helical turn)	Concentration of chloroquine required[c] (μg/ml)	Positive or negative supercoils[d]
>1000	0.000–−0.010	4	Positive
1000–400	−0,010–−0.026	10	Positive
400–250	−0.026–−0.042	30	Positive
325–215	−0.032–−0049	40	Positive
270–180	−0.039–−0.058	90	Positive
190–135	−0.054–−0.078	10	Negative

[a] Divide the length of the plasmid DNA (in bp) by the number of nucleosomes expected.
[b] The nucleosome density is related to the superhelix density by: $\sigma = -1/(\text{nucleosome density}/10.5)$.
[c] This is the concentration of chloroquine in the gel and running buffer which should give optimal resolution. Resolution is also greatly improved by a long, gel running time. These concentrations were determined empirically using a 3.25 kb plasmid.
[d] Topoisomers may migrate as positive or negative supercoils in chloroquine gels (see section 3.1 for further explanation).

concentrations required), starting with the $\Delta Lk = 0$ standard (see section 3.2.2). Therefore, it is important that the topoisomers in each standard overlap a little with the next standard in the series, such that bands can be identified (e.g. *Figure 1*). For larger plasmids, larger increments in ΔLk can be used because the number of topoisomers visible increases with the size of the plasmid. However, the maximum superhelix density will be the same because this does not depend on plasmid length. Thus, standards with superhelix densities of 0, − 0.013, − 0.026, − 0.039, − 0.052, and – 0.065 would be adequate for a plasmid of any size.

3.2.2 Relaxed DNA ($\Delta Lk = 0$)

The linking number of circular DNA depends on the salt concentration and on the temperature (11). Therefore, the definition of relaxed DNA has to be referred to particular conditions. This is not a serious problem because the dependence of ΔLk on temperature and salt concentration is mild. Ideally, $\Delta Lk = 0$ should refer to DNA relaxed under the same conditions used for relaxation of the reconstituted chromatin (0.2 M NaCl and 37 °C, as in ref. 1). All other standards can then be referred to this as $\Delta Lk = 0$. This point is important in accurate nucleosome counts and could otherwise lead to an error of 1 or 2 nucleosome cores (see *Figure 2*). Thus, for nucleosome counts, it is convenient to define $\Delta Lk = 0$ by the centre of the topoisomer distribution for DNA relaxed in 0.2 M NaCl at 37 °C.

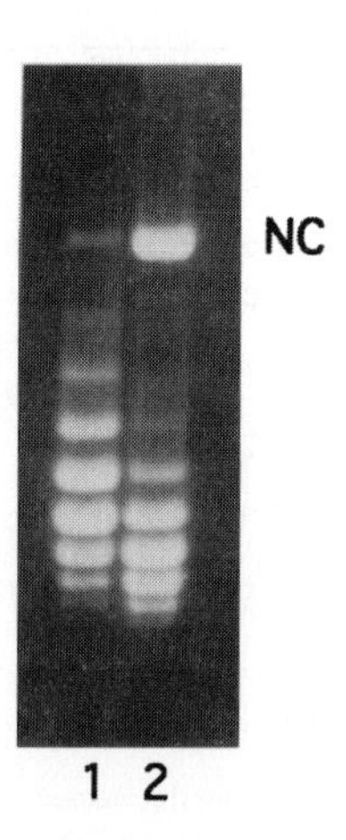

Figure 2. Dependence of DNA supercoiling on salt concentration and temperature. Plasmid pXP10 (3.25 kb) was relaxed with nicking–closing extract: Lane 1, in 0.2 M NaCl, 37 °C; lane 2, in 0.1 M NaCl, 20 °C. Samples were analysed in a gel containing chloroquine at 4 μg/ml. NC, nicked circle. The two samples differ by about 1.5 linking numbers (the sample in lane 2 is more positively supercoiled than that in lane 1).

Protocol 2. Preparation of topoisomer standards

Equipment and reagents

- Ethidium bromide (EtBr) (Aldrich; cat. no. 16,053–9) (**Caution**: mutagen)
- Topoisomerase I or nicking–closing extract
- Speed-Vac concentrator
- 10% (w/v) sodium dodecyl sulfate (SDS)
- 3 M sodium acetate (pH adjusted to 7.0 with acetic acid)
- Phenol/chloroform (1:1)
- Ethanol and 70% (v/v) aqueous ethanol
- Microcentrifuge
- 1-Butanol
- TE buffer: 10 mM Tris–HCl, pH 8.0, 1 mM Na-EDTA
- Dialysis buffer: 3 M NaCl in TE buffer (2 × 1 litre)
- Buffer 1: 0.2 M NaCl, 20 mM Tris–HCl, pH 8.0, 1 mM Na-EDTA, 5% (v/v) glycerol
- Water bath at 37 °C

Method

1. Prepare a stock solution of ethidium bromide at 2 mg/ml in water. This should be of high quality and accurately weighed out. Prepare a working solution of 0.1 mg/ml ethidium bromide for use in this experiment. Protect these solutions from light.
2. Determine the concentration of plasmid DNA spectrophotometrically using A_{260} (1 mg/ml) = 20. Adjust the concentration to 0.5–2 mg/ml. DNA can be stored in TE buffer at −20 °C indefinitely.
3. For each standard, dilute 10–100 μg of the plasmid to a final concentration of 0.1 mg/ml in buffer 1, and include ethidium bromide (using the 0.1 mg/ml stock solution) in amounts calculated using the following equation[a]:

   $$\mu g \text{ EtBr } \underline{\text{per } \mu g \text{ DNA}} \text{ to be added} = 0.9 \times \sigma = 0.9 \times (10.5/PL) \times \Delta Lk \quad [3]$$

   where *PL* is the length of the plasmid in base pairs. This should yield plasmids with ΔLk values within 2 linking numbers of the expected value.
5. Add 1 unit nicking–closing extract or topoisomerase I per μg DNA and protect the tubes from light to prevent nicking of DNA.
6. Incubate at 37 °C for 2 h.
7. Stop the reaction by adding SDS to a final concentration of 0.2% .
8. Add one-tenth volume 3 M sodium acetate (pH 7.0).
9. Extract several times with 4 vols 1-butanol to remove most of the ethidium, adjusting the volume with water each time.
10. Extract with an equal volume of phenol/chloroform (1:1) to remove protein and then chloroform to remove phenol.
11. Dialyse in the dark at 4°C against 3 M NaCl in TE buffer to remove residual ethidium (this is important because trace amounts of ethidium will cause problems), changing the buffer once.

Protocol 2. *Continued*

12. Dialyse against TE buffer at 4°C to remove NaCl.
13. Inspect the standards in chloroquine gels (see below).

[a] This equation was derived assuming that each ethidium molecule bound unwinds the DNA helix by 24° (7, 17) and that all the ethidium is bound to DNA.

3.2.3 Analysis of standards in a series of gels at different chloroquine concentrations

The topoisomers in each standard must be identified by analysis in a series of chloroquine gels, using the centre of the topoisomer distribution in the standard relaxed in the absence of ethidium as $\Delta Lk = 0$ (use *Table 1* as a guide).

3.3 Electrophoresis in chloroquine gels

Protocol 3. Counting nucleosomes in chloroquine gels[a]

Equipment and reagents

- Slab gel apparatus (Bio-Rad: the large 'sub-cell DNA electrophoresis cell' or the smaller 'mini-sub cell') with a 15 cm × 15 cm tray and a 20- or 30-slot comb for the large apparatus, or a 10 × 7 cm tray and an 8-slot comb for the mini-gel apparatus.
- 10 × TPE: 400 mM Tris-base (48.44 g per litre), 300 mM sodium phosphate (41.4 g $NaH_2PO_4 \cdot H_2O$ per litre), 10 mM sodium-EDTA (3.72 g per litre); the final pH should be about 8.2 (without adjustment)
- Running buffer (1 × TPE (18)) containing chloroquine at the appropriate concentration (1.6 litres for the large apparatus; 250 ml for the mini-gel)
- Power supply
- Peristaltic pump (Pharmacia Biotech) capable of recirculating the running buffer at a rate of at least 250 ml/h.
- Chloroquine diphosphate (Sigma; C-6628). Chloroquine is light-sensitive and a fresh solution at 50 mg/ml should be prepared on the day of use.
- Agarose (Ultrapure grade from Gibco/BRL).
- A cover for the gel apparatus to keep it dark (to prevent degradation of chloroquine).

Method

1. Make an appropriate choice of chloroquine concentration from a rough estimate of the nucleosome density of the sample, i.e. the number of base pairs per nucleosome core. For chromatin reconstituted using purified core histones *in vitro*, the nucleosome density should be close to that expected from the histone:DNA ratio used, because the reaction is stoichiometric. Chromatin assembled in extracts is likely to contain nucleosomes at close to physiological spacing. Estimate the appropriate chloroquine concentration from the nucleosome density using *Table 1*.
2. The choice of agarose concentration depends on the size of the plasmid to be analysed. Use a 1.2% (w/v) gel for resolving topo-

isomers of plasmids about 3 kb in size. Use a 1% gel for plasmids of about 4.5 kb. For larger plasmids (e.g. 6 or 7 kb), use a lower agarose concentration (0.7% or 0.8%). For large gels (15 x15 cm tray), heat the agarose in 120 ml 1 × TPE buffer to dissolve. For small gels (10 × 7 cm tray), use 40 ml 1 × TPE. After boiling the agarose and allowing it to cool a little, add chloroquine to the desired concentration.

3. Prepare samples and appropriate standards (section 3.2) in sample buffer, with or without chloroquine. If chloroquine is included omit Bromophenol blue, which can form a precipitate with chloroquine; include the dye in unused tracks.
4. Run large gels slowly (overnight) to maximize resolution. The best running conditions depend on the DNA size and the agarose concentration. As a guide, for a 3.25 kb plasmid in a 1.2% gel, 38 V for 15 h maximizes resolution without topoisomers migrating off the end of the gel. For a 4.5 kb plasmid, use a 1% gel and the same electrophoresis conditions. For larger plasmids, *either* run the gel for much longer *or*, better, reduce the agarose concentration. Cover the gel box to reduce exposure of the chloroquine to light. For a 3.25 kb plasmid, run 1.2% mini-gels at 40 V for 5–6 h.
5. Recirculate the running buffer at an adequate rate during electrophoresis (250 ml/h for large gels).

[a] If the amount of sample DNA available is limited and cannot easily be detected by staining, then an alternative method is visualization by Southern blot hybridization. If labelled DNA is used, then simple autoradiography is possible (e.g. in experiments coupling DNA replication with chromatin assembly (19).

3.4 Detecting the DNA

Protocol 4. Detecting the DNA with ethidium bromide[a]

Reagents

- 1 × TPE buffer (*Protocol 1*)
- UV light box
- 1 μg/ml ethidium bromide (see *Protocol 2*) in 1 × TPE buffer

Method

1. Chloroquine inhibits the binding of ethidium (they are both intercalating agents) and therefore if the concentration of chloroquine in the gel is relatively high (over 10 μg/ml), wash the gel in one or two changes of 1 × TPE buffer for 30–60 min on a rocking platform. If the chloroquine concentration in the gel is very high (above 100 μg/ml), wash several times for optimal staining.

Protocol 4. *Continued*

2. After the washes, stain with 1 μg/ml ethidium bromide in 1 × TPE buffer for 1 hour.
3. Destain the gel (if necessary) by washing once more with buffer, and photograph it on a UV light box.

[a] See footnote to *Protocol 3*.

3.5 Determination of the average number of nucleosome cores

A typical analysis is shown in *Figures 3* and *4*. Samples derived from a set of reconstitutes, relaxed using nicking–closing extract, were analysed in a gel containing chloroquine at 40 μg/ml. Visual inspection of *Figure 3* indicates that ΔLk can be determined for all the reconstituted chromatin samples because, with the possible exception of sample 5, the topoisomer distributions are clearly resolved. The nucleosome-free plasmid relaxed as a control (C) is not resolved at this chloroquine concentration. The use of topoisomerase I results in a small increase in the amount of nicked circle, which might reflect the fact that the relaxation mechanism involves the nicking and re-closing of DNA.

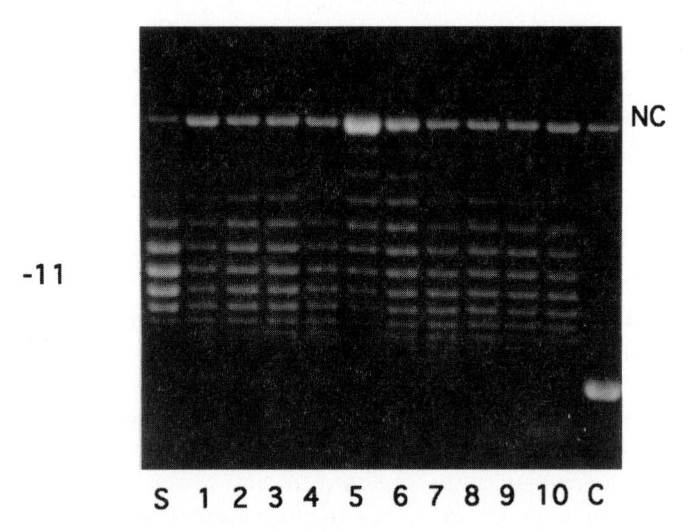

Figure 3. Counting nucleosomes using the topoisomerase method: analysis of topoisomer distribution in a chloroquine gel. Nucleosome cores were reconstituted on plasmid pXP10 (3.25 kb) by salt/urea dialysis (4) and prepared for analysis as described in *Protocol 1*. Topoisomers were analysed in a gel containing chloroquine at 40 μg/ml. S, standard with average $\Delta Lk = -11$. Lanes 1–10, plasmid DNA extracted from different reconstituted chromatin samples after relaxation. C, control (nucleosome-free pXP10 after relaxation under the same conditions as the reconstitutes). NC, nicked circle.

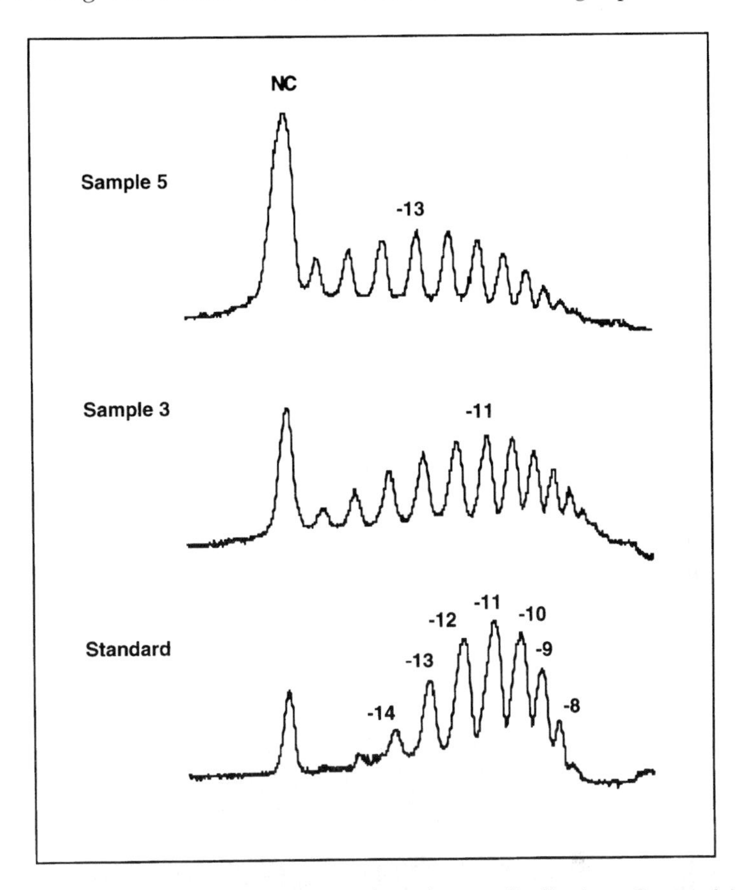

Figure 4. Examples of densitometer scans of topoisomer distributions. Scans of the standard and two samples (3 and 5) from a Polaroid negative of the gel shown in *Figure 3*. NC, nicked circle. Values of ΔLk assigned using the standard are indicated. For quantitation, a flat baseline was drawn and then the peak heights were measured using the software associated with the densitometer (section 3.5). The highest peak marks the centre of the distribution, but if the two central peaks are of equal height, then ΔLk is reported as half-integral. The scans are shown unsmoothed. Note that the topoisomer distribution in the nucleosomal samples is broader than in the standard.

The average number of nucleosome cores in each reconstitute is determined by comparison with the standard (in this case, $\Delta Lk = -11$), often by visual inspection of the gel. However, the formation of nucleosome cores on a plasmid broadens the topoisomer distribution of the plasmid, relative to nucleosome-free plasmid (*Figures 3* and *4*), and it can be difficult to identify the central band (strongest intensity) unambiguously. This problem can be circumvented by densitometry. The negative of a Polaroid photograph of the gel is scanned (*Figure 4*) to determine the centre of the distribution (the strongest band) which is equal to the mean value of ΔLk. More details are given in the legend to *Figure 4*. We use a Hoeffer transmittance/reflectance scanning densitometer (Model GS300) with its associated software (the GS360 data system). The distribution of topoisomers is quantitatively symmetrical and follows the Boltzmann distribution (7, 20). The mean value of ΔLk is equal to the average number of nucleosome cores.

After use of a scanning densitometer (see *Figure 4*) and comparison with the linking number standard, the samples were determined to have the following average numbers of nucleosome cores: samples 1, 2, 3, and 9: 11 nucleosome cores; samples 4, 8, and 10: 10.5 nucleosome cores; sample 5: 13 nucleosome cores; sample 6: 11.5 nucleosome cores; sample 7: 10 nucleosome cores.

Acknowledgements

I thank Mary O'Dea and Martin Gellert for provision of the nicking–closing extract and for much advice. I thank Rodolfo Ghirlando for useful comments on the manuscript.

References

1. Simpson, R. T., Thoma, F., and Brubaker, J. M. (1985). *Cell*, **42**, 799.
2. Norton, V. G., Imai, B. S., Yau, P., and Bradbury, E. M. (1989). *Cell*, **57**, 449.
3. Norton, V. G., Marvin, K. W., Yau, P., and Bradbury, E. M. (1990). *J. Biol. Chem.*, **265**, 19848.
4. Clark, D. J. and Felsenfeld, G. (1991). *EMBO J.*, **10**, 387.
5. Clark, D. J., Ghirlando, R., Felsenfeld, G., and Eisenberg, H. (1993). *J. Mol. Biol.*, **234**, 297.
6. Camerini-Otero, R. D. and Felsenfeld, G. (1977). *Nucl. Acids Res.*, **4**, 1159.
7. Keller, W. (1975). *Proc. Natl Acad. Sci. USA*, **72**, 4876.
8. Lutter, L. C., Judis, L., and Paretti, R. F. (1992). *Mol. Cell. Biol.*, **12**, 5004.
9. Pederson, D. S. and Morse, R. H. (1990). *EMBO J.*, **9**, 1873.
10. Crick, F. H. C. (1976). *Proc. Natl Acad. Sci. USA*, **73**, 2639.
11. Depew, R. E. and Wang, J. C. (1975). *Proc. Natl Acad. Sci. USA*, **72**, 4275.
12. Richmond, T. J., Finch, J. T., Rushton, B., Rhodes, D., and Klug, A. (1984). *Nature*, **311**, 532.
13. White, J. H. and Bauer, W. R. (1989). *Cell*, **56**, 9.
14. Hayes, J. J., Tullius, T. D., and Wolffe, A. P. (1990). *Proc. Natl Acad. Sci. USA*, **87**, 7405.
15. Almouzni, G., Mechali, M., and Wolffe, A. P. (1990). *EMBO J.*, **9**, 573.
16. Kamakaka, R. T., Bulger, M., and Kadonaga, J. T. (1993). *Genes Dev.*, **7**, 1779.
17. Shure, M. and Vinograd, J. (1976). *Cell*, **8**, 215.
18. Garner, M. M., Felsenfeld, G., O'Dea, M. H., and Gellert, M. (1987). *Proc. Natl Acad. Sci. USA*, **84**, 2620.
19. Almouzni, G. and Mechali, M. (1988). *EMBO J.*, **7**, 4355.
20. Pulleyblank, D. E., Shuer, M., Tang, D., Vinograd, J., and Vosberg, H. (1975). *Proc. Natl Acad. Sci. USA*, **72**, 4280.

8

In vitro reconstitution and analysis of nucleosome positioning

COLIN DAVEY, SARI PENNINGS, and JAMES ALLAN

1. Introduction

It is now established that positioned nucleosomes play a role in controlling the genetic functions of DNA (1–3). This is a consequence of the ability of the core-histone octamer to: (a) sequester a DNA sequence, by incorporating it within the nucleosome; (b) ensure that a DNA sequence remains exposed within the linker DNA between nucleosomes; or (c) to bring distant DNA sequences into close proximity. The precise positioning of nucleosomes with respect to the underlying DNA sequence thus determines its capacity to interact with, or to promote interaction between, other DNA-binding factors. The specific disposition of nucleosomes over long stretches of DNA is also likely to determine the architecture and stability of the higher-order chromatin fibre, which may, in turn, dictate its functional properties.

Although various factors can contribute to the positioning of nucleosomes (2), the DNA sequence itself is a fundamental determinant (1, 2, 4–6). A favourable distribution of short sequence elements is thought to dictate compatibility with the energetically unfavourable bending required to wrap the DNA around the core-histone octamer (1, 2). The DNA also appears to be substantially deformed at symmetrically located sites close to the dyad axis of the core particle (7, 8), distortions which may be more favourably accommodated by certain DNA sequences.

Nucleosome mapping *in vitro* is carried out after the reconstitution of purified core histones onto DNA fragments. Because only histones and DNA are involved, the positions adopted specifically reflect the propensity of particular DNA sequences to accommodate the histone octamer. In many instances, the nucleosome positioning sites that are detected *in vivo* are the same as those adopted upon *in vitro* reconstitution (2, 3, 9).

Here we describe the methods used to prepare reconstituted chromatin and outline the techniques most frequently used to determine the positions adopted by the core-histone octamer on the DNA. These procedures provide information about the rotational setting of the DNA on the surface of a histone octamer or the translational positioning of octamers along a DNA sequence. Furthermore, some of the methods can provide quantitative information concerning the relative binding strengths of different positioning sites.

2. Reconstitution of chromatin

The most commonly employed methods for reconstituting chromatin involve: (a) dialysis or dilution from high-salt conditions after the addition of purified core histones to DNA; or (b) the redistribution of core histones onto DNA from a donor chromatin. Methods which employ nuclear assembly extracts (10) or histone chaperones (11) are not dealt with here. The preparation of purified core-histone protein and donor chromatins (mono- or polynucleosomes) have been described in Chapters 1 and 2. Donor chromatin should be stripped of non-core histone proteins, usually by salt extraction (12, 13). The DNA for reconstitution may be in the form of a short fragment, designed to accommodate a single histone octamer, or a plasmid DNA molecule capable of binding numerous octamers; this will be determined by the type of positioning information required and is therefore dealt with in the context of each method used to assess positioning.

2.1 Reconstitution with purified core histones

Both short DNA fragments and plasmids can be reconstituted with purified core histones by dialysis from high-salt conditions. Short fragments that are available only in small quantities should be labelled and supplemented with carrier DNA to enable the histone to DNA ratio to be established for reconstitution. Dilute nucleosome samples can dissociate and it is important to maintain a chromatin concentration (10 μg/ml or above) where this is minimized. Control of the histone to DNA ratio during reconstitution is important because DNA fragments ($\geq$250 bp) are able to accommodate more than one octamer and this will be enhanced at a high histone to DNA ratio. Conversely, a low ratio will leave much of the DNA unreconstituted. Reconstitution efficiency also depends upon the affinity of core histones for the DNA sequence under examination (14). It is advisable to determine the optimal histone to DNA ratio experimentally, assessing the extent of reconstitution on nucleoprotein gels (see below). Ultimately, it may be necessary to purify reconstituted monomers from dimers and free DNA by sucrose gradient sedimentation (13, 15, and Chapter 1).

With regard to the mapping techniques described below, it is usually appropriate to reconstitute plasmids at a relatively low histone to DNA ratio ($\approx$ 0.5:1 (w/w)). This will reduce the likelihood of octamer–octamer interactions influencing positioning and the possibility of histone octamers sliding together and aggregating during processing of the reconstituted chromatin.

High-salt reconstitution involves adding DNA and histones together under conditions where they are dissociated, and then gradually reducing the salt concentration to effect their association. In our experience, the pathway by which this is achieved does not substantially influence the positions adopted by core-histone octamers. For example, nucleosomes reconstituted onto the chicken β-globin promoter adopt the same positions after: (a) stepwise dialysis from 2 M NaCl, 5 M urea to low-salt conditions ($\approx$ 10 mM) over a period in excess of 48 hours; or (b) gradient dialysis from 2 M NaCl to low-salt conditions over 6 hours (16). Some methods simply involve a stepwise dilution from 2 M NaCl over only a few hours (17). A method for high-salt reconstitution using purified core histones is given in *Protocol 1*.

Protocol 1. Chromatin reconstitution with purified core histones

Equipment and reagents

- Core histones (Chapter 2)
- DNA fragment or plasmid
- TEP buffer: 10 mM Tris–HCl, pH 7.5, 0.1 mM EDTA, 0.1 mM PMSF
- Carrier DNA (optional)
- Microdialysis system for small volume dialysis (Gibco/BRL)

Method

1. Mix 20–50 μg of DNA ((fragment + carrier) or plasmid) with core histones to give a histone to DNA ratio (w/w) of 1:1 (fragment) or 0.5:1 (plasmid) in TEP buffer containing 2.0 M NaCl to give a final DNA concentration of 0.1 mg DNA/ml.
2. Dialyse against a 1 litre, linear 2.0 M–0.4 M NaCl gradient (in TEP buffer) at 4°C for 3–4 h.
3. Dialyse against low salt (16 mM NaCl in TEP buffer) at 4°C for 3–4 h or overnight[a].

[a] This final buffer can be modified to accommodate the requirements of subsequent analyses.

2.2 Chromatin reconstitution by core-histone redistribution from donor chromatin

This procedure is suitable for the reconstitution of short DNA fragments. The DNA is incubated with a large excess of stripped donor chromatin under high-salt conditions, which encourages the redistribution of octamers from the donor to the target DNA, followed by dialysis to low-salt concentrations (13, 15, 18). This procedure has a number of advantages. First, the histone to DNA ratio is not of major concern because it is determined by the large excess of donor chromatin, and, in principle, it is difficult to over-titrate the target DNA. However, some DNA fragments may have an especially high affinity for core histones (14) and at equilibrium may end up with a greater than average octamer density. Again, optimal reconstitution conditions should be determined experimentally. Second, the donor chromatin serves as a carrier, maintaining a suitably high chromatin concentration. Finally, if long donor chromatin fragments are used for transfer, they can be separated from the reconstituted fragment by sucrose gradient centrifugation (19); however, the reconstitute will become extensively diluted during centrifugation and may be more suceptible to dissociation. A method for reconstitution by core-histone redistribution is given in *Protocol 2*.

Protocol 2. Reconstitution by core-histone redistribution

Equipment and reagents

- Linker histone-depleted chromatin (Chapter 1)
- DNA fragment
- TEP buffer (*Protocol 1*)
- Microdialysis system (as *Protocol 1*)
- 1.2 M, 0.6 M, 16 mM NaCl in TEP buffer

Method

1. Mix 0.1–1.0 μg of fragment DNA with linker histone-depleted donor chromatin to give a chromatin to DNA ratio (w/w) of between 50:1–100:1 in 200 μl TEP buffer, 1.2 M NaCl. Incubate at 37 °C for 20 min.
2. Dialyse against 0.6 M NaCl (in TEP buffer) at 4 °C for 3–4 h.
3. Dialyse against 16 mM NaCl (in TEP buffer) at 4°C for 3–4 h or overnight[a].

[a] Nucleosome monomers may be purified by sedimentation through a 6–40% isokinetic sucrose gradient containing 16 mM NaCl (in TEP buffer). Centrifuge at 4°C for 17 h at 286 000 g (41 000 r.p.m.) in a Beckman SW40.1 rotor. Remove sucrose from pooled fractions by dialysis against 16 mM NaCl (in TEP buffer) and store at 4°C.

3. Preparation of core-particle DNA

A number of methods for mapping nucleosome positioning involve the preparation of core-particle DNA from reconstituted chromatin. This population of molecules should represent, both qualitatively and quantitatively, the positions adopted by histone octamers assembled onto the DNA. The digestion of chromatin with micrococcal nuclease (MNase) can induce sliding of histone octamers and may selectively deplete the population of core particles susceptible to internal cleavage. The enzyme also displays sequence specificity for A+T-rich DNA and this may, with respect to the DNA ends, result in a biased core-particle population. Consequently, one should employ MNase-digestion conditions designed to alleviate these problems.

Although homogeneous preparations of core-particle DNA can be obtained by digestion of chromatin at 37 °C (12), we recommend that 0 °C be employed for the initial stages of digestion. This will increase nucleosome stability and reduce the exonuclease activity of the enzyme, which promotes both sliding and processive invasion at the termini of core particles. Nucleosomes prepared at 0 °C can be briefly heated to 37 °C at the end of the reaction to ensure trimming to the core-particle boundary. Alternatively, nucleosomes prepared by a brief exposure to micrococcal nuclease can be trimmed to core particles by the combined use of exonuclease III and S1 nuclease (20). It is strongly recommended that test digests, monitored by gel electrophoresis, are undertaken to determine the optimal digestion conditions, the extent of sliding, and the recovery of core-particle DNA.

DNA isolated from core particles should be purified by preparative electrophoresis in an agarose or polyacrylamide gel to ensure that the final sample comprises only ≈ 146 bp fragments. A method for the preparation of core-particle DNA from reconstituted plasmid is given in *Protocol 3*.

Protocol 3. Preparation of core-particle DNA

Equipment and reagents

- Reconstituted chromatin (*Protocol 1*)
- 1000 units/ml micrococcal nuclease (Sigma)
- TEP buffer (*Protocol 1*)
- Buffered phenol:chloroform:IAA
- Chloroform:IAA (*Protocol 5*)
- 7% polyacrylamide or 1.8% agarose gel
- Dialysis bag[a]

Method

1. Digest reconstituted chromatin, at a concentration of 80–120 μg/ml in TEP buffer containing 16 mM NaCl and supplemented with 1 mM $CaCl_2$, with micrococcal nuclease (20 units/ml) for 30–40 min on ice.
2. Transfer the sample to 37 °C for 3 min. Stop the reaction by adding EDTA to 5 mM.
3. Purify DNA from reactions by extraction with phenol and chloroform and precipitate with ethanol. Dissolve samples in the appropriate gel loading buffer.
4. Fractionate the DNA in a native 7% polyacrylamide or 1.8% agarose gel. Excise the core-particle DNA (≈ 146 bp) as a gel slice and recover by electroelution in a dialysis bag[a]. Purify and precipitate with ethanol.

[a] See ref. 38.

4. Nucleoprotein gel analysis of positioned nucleosomes

Most methods for analysing the positioning of nucleosomes on DNA sequences are disruptive. They typically involve invasive enzymatic procedures, followed by gel electrophoretic analysis to establish the cutting patterns on the nucleosomal DNA. These procedures, which are outlined below, have revealed that on most DNA sequences, histone octamers can bind to a number of, often overlapping, positions. Nucleoprotein gel electrophoresis constitutes a powerful tool to analyse and manipulate populations of variously positioned nucleosomes (positioning isomers) in their native state (21), and constitutes the only technique available for their fractionation (*Protocol 4*). This simple method yields a direct and visual estimate of the number of different positions adopted by the core-histone octamer as well as their locations relative to the middle of the DNA fragment. Because the nucleosomes are left intact by the separation according to position, they are suitable for further experimentation (21, 22).

The separation of positioned nucleosomes by polyacrylamide gel electrophoresis seems to exploit the same mechanism which makes this technique suitable for identifying bent DNA and DNA-binding proteins (23). In the latter case, the association of a protein with a DNA fragment causes its electrophoretic retardation, referred to as a bandshift. This is caused by the additional molecular weight and may be further enhanced by protein-induced

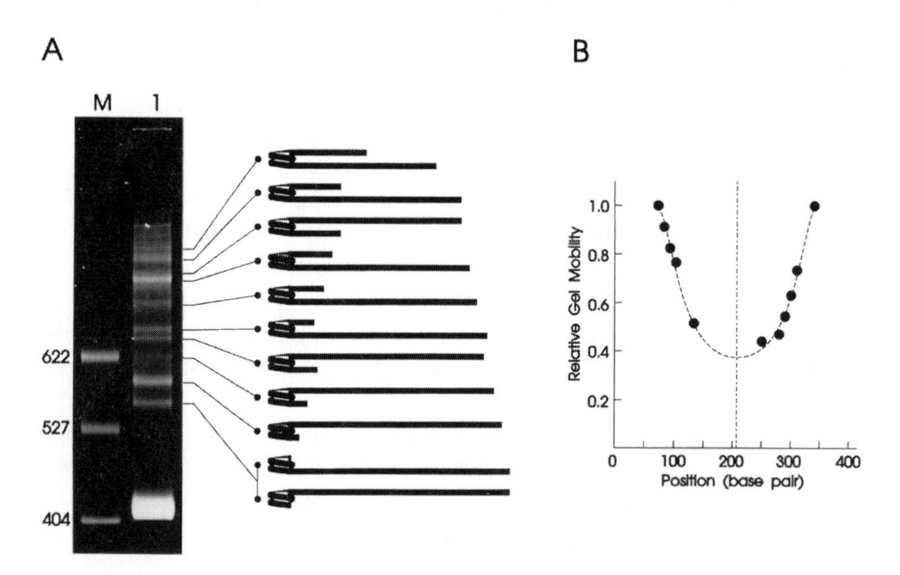

Figure 1. (A) Nucleoprotein polyacrylamide gel analysis of mononucleosomes reconstituted onto a 207 bp dimer (414 bp) of sea-urchin 5S rDNA showing separation according to nucleosome position (Lane 1). DNA marker (M) fragment lengths are indicated (bp). To the side are drawings of the proposed positioned mononucleosomes that give rise to the observed bands (protruding DNA and nucleosome dimensions are drawn to scale). Shaded octamers represent the dominant position (21). (B) Graph depicting the correlation between gel electrophoretic mobility of the 5S rDNA mononucleosomes and their position (taken from the midpoint of the core particle) on the DNA fragment.

bending of the DNA fragment. Nucleosomes form no exception to this rule, as they can be distinguished from naked DNA by their much slower migration. DNA fragments longer than about 190 bp typically shift to a series of bands upon reconstitution, which represent discrete nucleosome positions (21, 24) (*Figure 1A*). The electrophoretic mobility of these positioning isomers is a function of the proximity of the histone octamer to the centre of the DNA fragment (21, 25). Maximal retardation is found for nucleosomes located centrally on the DNA (*Figure 1B*), similar to the behaviour found for bent DNA. In analogy with protein-induced DNA bending, the angle between the DNA entering and leaving the core particle seems to be responsible for this property. This conformational parameter determines the electrophoretic mobility of nucleosomes in polyacrylamide gels in addition to particle mass and charge (26). In less homogenous samples, band heterogeneity caused by differences in histone content or DNA length can mask the separation according to nucleosome position. Glycerol, which is often included in polyacrylamide gels to stabilize protein–DNA complexes, reduces conformational separation and must therefore be omitted (26).

Protocol 4. Nucleoprotein gel analysis of positioned nucleosomes

Equipment and reagents

- Reconstituted mononucleosomes (DNA must be longer than 190 bp)
- Vertical slab gel apparatus for 17 cm × 20 cm (approx.) 5% polyacrylamide gels
- 0.5 × TBE buffer: 45 mM Tris-base, 45 mM boric acid, 1.25 mM EDTA, pH 8.3
- 10 mM NaCl, 0.2 mM EDTA, 10 mM Tris–HCl, pH 7.4
- Ficoll 400
- Bromophenol blue
- Xylene cyanol dye
- Ethidium bromide

Method

1. Cast a 5% polyacrylamide (29:1 acrylamide to bisacrylamide) vertical slab gel in 0.5 × TBE buffer.
2. Pre-run the gel at 4°C[a] for at least 1 h, at 10 V/cm in 0.5 × TBE buffer.
3. Add loading buffer to nucleosomes[b] by vortexing into the samples, to a final concentration of 0.5 × TBE, 3.5% Ficoll 400[c], 0.05% Bromophenol blue, and 0.05% xylene cyanol dye.
4. Add the nucleosome samples into the wells, and run the gel at 4 °C for about 1000 volt hours (for nucleosomal DNA ranging between 200 and 400 bp) at a maximum of 10 V/cm.
5. Stain the gel[d] in 2 µg/ml ethidium bromide for 20 min, destain in H_2O, and photograph. Nucleosomes assembled on radiolabelled DNA are analysed after drying and autoradiography of the gel.

Protocol 4. *Continued*

6. If populations of positioned nucleosomes are to be processed further, keep the samples on ice. Include 2 mM $MgCl_2$ or $CaCl_2$[e] to prevent dynamic redistribution of nucleosomes between positions[f].

[a] It is imperative to run the gel in the cold.
[b] Use 0.5–2 μg of nucleosomes per lane, depending on the size of the well and the number of expected bands. Ensure nucleosomes are at a concentration of at least 50 μg/ml in a 10 mM NaCl, 0.2 mM EDTA, 10 mM Tris–HCl (pH 7.4) buffer.
[c] High salt, glycerol, or sucrose reduce the resolution of the bands.
[d] If the separated nucleosomes are to be processed further, a duplicate lane can be included for this purpose. **Caution**—EthBr is a powerful mutagen.
[e] See ref. 24.
[f] See ref. 21.

5. Nucleosome mapping techniques

The methods used to resolve the positions occupied by core-histone octamers on DNA can be divided into two categories. Some techniques are appropriate for determining the precise rotational setting of the DNA on the surface of the octamer, but do not unambiguously establish the boundaries of the core particle. Conversely, many methods have been described to map the boundaries of the core particle and thereby its translational positioning, but provide, at best, only an approximation of the rotational setting.

5.1 Mapping rotational setting

The orientation of the DNA helix on the histone surface can be deduced by cleaving the DNA in reconstituted nucleosomes, either enzymatically with DNase I or chemically with hydroxyl radicals, and subsequently determining the locations of the cleavage sites by electrophoresis in a denaturing polyacrylamide gel (Chapter 3 and refs 8, 17). Because the reagents are prevented from attacking the face of the DNA associated with the histone core, cleavage occurs predominantly on the exposed surface and is therefore modulated to reflect the helical periodicity of the DNA. DNase I analysis, although extensively used (4, 8, 13), has recently been superseded by hydroxyl-radical footprinting (8, 27) because the latter reagent displays little of the sequence specificity associated with the endonuclease.

For DNase I or hydroxyl-radical footprinting, reconstitute nucleosomes onto a DNA fragment uniquely end-labelled with either the Klenow fragment of DNA polymerase I (3′) or T4 polynucleotide kinase (5′). Relatively large DNA fragments can be used (17), but shorter fragments, which bind only a single octamer, provide data that are simpler to interpret. It is essential that cleavage reactions are carried out on both histone-reconstituted and protein-free DNA to correct for any inherent modulation in cleavage associated with the DNA sequence. Nucleotide-specific chemical cleavage of the end-labelled DNA provides the most appropriate size marker.

The chromatin substrate should not contain excessive unreconstituted DNA. Sucrose gradient fractionation can be employed to prepare pure mono-nucleosomes, but this will not resolve positioning isomers. To overcome this problem, chromatin samples can be subjected to cleavage and then fractionated by electrophoresis in a native polyacrylamide gel (*Protocol 4*). DNA in bands corresponding to the different positioning isomers can be recovered from the gel and analysed in denaturing gels (24).

A variety of conditions can be used to digest reconstituted chromatin with DNase I (8, 13, 17). We routinely carry out reactions in TEP buffer (*Protocol 1*) supplemented with 1 mM $MgCl_2$, and digest with DNase I (0.5 units/ml) at 20°C over a time course of 10 minutes. Digests of protein-free DNA require less enzyme or a shorter time course. In either case, the chromatin concentration should be at least 10 μg/ml, which may necessitate the addition of unlabelled carrier chromatin. Stop the reactions by the addition of EDTA (5 mM) and SDS (0.2%). Detailed descriptions of the more demanding methods and conditions for hydroxyl-radical cleavage have been presented previously (27). After DNase I or hydroxyl-radical cleavage, purify DNA by extraction with phenol and chloroform and precipitate with ethanol. Dissolve samples in gel loading buffer, denature at 95°C and electrophorese in a 6% denaturing polyacrylamide gel.

For both of the above procedures, optimal conditions are those that effect an average of a single cleavage per DNA molecule. This should be determined experimentally and is achieved when approximately 50% of the fragment remains undigested.

5.2 Mapping translational positioning

When core-histone octamers are reconstituted onto DNA they often adopt a variety of locations with respect to the sequence, termed their translational positions. The methods employed to define these positioning sites invariably involve determining the boundaries of the 146 bp of DNA that is tightly associated with the histone octamer. The direct approach uses exonuclease III. The other methods involve purifying DNA from core particles, prepared from reconstituted chromatin, the ends of which are mapped with respect to the sequence used for reconstitution.

5.2.1 Mapping with exonuclease III

Exonuclease III (Exo III) digests double-stranded DNA in a 3′ → 5′ direction from a 3′-OH end. The progress of this enzyme is usually strongly impeded at the boundary of a core particle producing a distinct, chromatin-specific pause site during the course of digestion. If the substrate is 5′ end-labelled, the location of this pause site, and therefore the boundary of the nucleosome, can be determined by measuring the length of the Exo III-resistant products (15, 28).

Reconstitute nucleosomes onto a DNA fragment labelled at only one 5′ end. Typically, both ends of a restriction fragment are labelled with T4 polynucleotide kinase and then one is removed with a restriction enzyme. Avoid restriction enzymes which generate a protruding 3′ terminus, as this structure is a very poor substrate for Exo III. The final DNA fragment should be 250–300 bp, long enough to accomodate an octamer and to provide DNA, extending from this structure, which Exo III can degrade to a product short enough to be distinguished from the starting material. To ensure that only monomer nucleosomes are subjected to digestion, these are often purified by sucrose gradient centrifugation (13, 15).

Exo III digestion is carried out at a chromatin concentration of at least 10 μg/ml, in TEP buffer (*Protocol 1*) supplemented with 1 mM $MgCl_2$. Digestion should proceed for 15 minutes at 37 °C using a range of enzyme concentrations (50–400 units/ml). Stop reactions by the addition of EDTA and SDS, purify the DNA, and ethanol-precipitate. Dissolve samples in gel loading buffer, denature at 95 °C, and electrophorese in a 6% denaturing polyacrylamide gel. Again it is essential to analyse reactions carried out on protein-free DNA because, even on this substrate, Exo III displays transient pausing.

Figure 2 shows an Exo III analysis of a nucleosomal substrate. In the mononucleosome digest a set of four prominent bands, absent from the DNA digest, reflect the nucleosome boundary. The largest band corresponds to the actual boundary of the core particle whereas the others, 10, 20, and 30 nucleotides shorter, reflect the capacity of Exo III to invade the nucleosome, pausing at each helical turn of the DNA. From the size of the largest band, the boundary can be mapped with respect to the labelled 5′ end. To map the other boundary, the experiment was repeated with the same DNA fragment labelled at the other 5′ end.

Exo III analysis is most appropriate for mapping short DNA fragments, particularly if the octamer is preferentially bound at a unique location. If the DNA contains numerous positioning sites, the reconstitute will be heterogeneous in positioning isomers, which will complicate Exo III analysis, particularly if these share the same rotational setting. Exo III analysis is not suited to providing information concerning the relative binding strengths (frequency of occupation) of different sites. This is because the intensity of the pause in Exo III digestion partly reflects the sequence of the terminal DNA segments of the core particle and the strength of their association with the histone octamer. These sequences are not instrumental in determining positioning (2) since strong nucleosome positioning sites can present weak boundaries to Exo III (15).

Most of the remaining methods we describe involve, as a starting point, the preparation of core-particle DNA from MNase-digested chromatin (*Protocol 3*).

5.2.2 Mapping with restriction enzymes

Restriction enzymes can be employed in two ways to investigate the location of nucleosomes on DNA sequences. The first approach is based on the

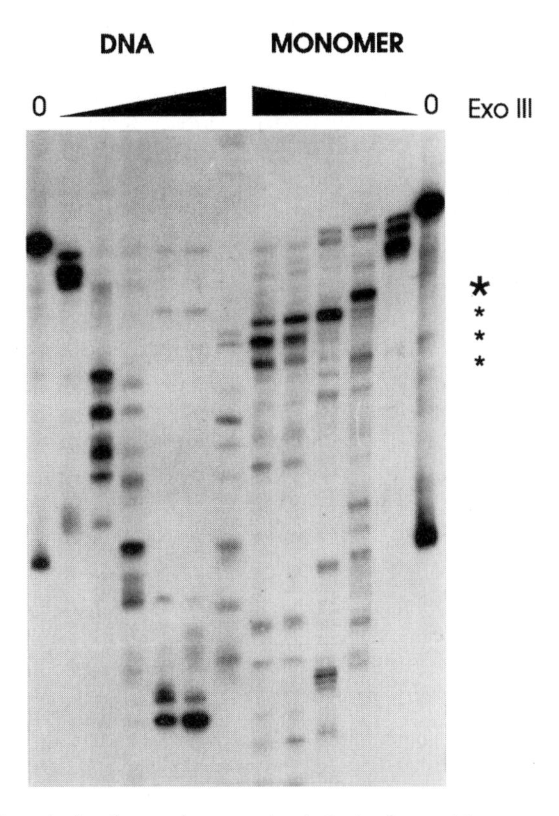

Figure 2. Exo III analysis of a nucleosomal substrate, formed by reconstituting histone octamers onto a 310 bp fragment of the chicken β-globin gene promoter. Samples were digested with increasing concentrations of Exo III and analysed on a 6% denaturing polyacrylamide gel. The stars indicate the chromatin-specific, Exo III pause sites.

premise that restriction sites within a core particle are inaccessible to cutting, while those in the linker DNA are accessible. Assaying restriction endonuclease accessibility on chromatin substrates provides low-resolution positioning information and can be used as a quick method to assess the presence or absence of a nucleosome on a particular DNA sequence (29). The approach has mainly been used to monitor changes in local chromatin structure associated with gene activation *in vivo* (9, 30).

If the region being mapped contains one or more restriction enzyme sites unique to the fragment or plasmid DNA used for reconstitution, these enzymes can be used as a short-range probe to locate the boundaries of core-particle DNA fragments which incorporate the site (16, 20, 31). Digestion of core DNA will produce a pair of fragments (for each positioning isomer), whose sizes add up to ≈ 146 bp, and define the location of the core-particle boundaries with respect to the restriction site. At least two restriction

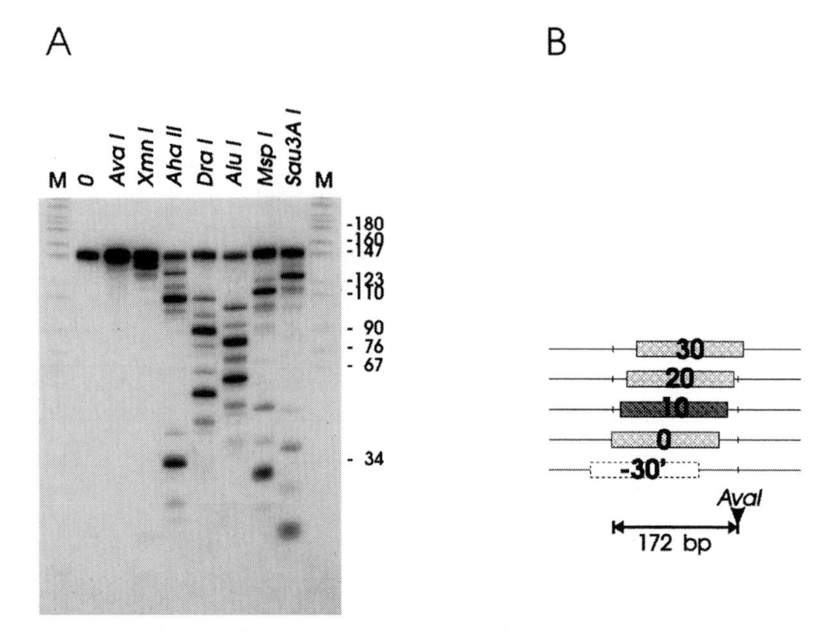

Figure 3. Restriction enzyme mapping of core-particle positioning in chromatin reconstituted on tandemly repeated, 172 bp, sea-urchin 5S rDNA. (A) Denaturing 6% polyacrylamide gel of the fragments of 5′ end-labelled, core-particle DNA, produced by restriction with the enzymes indicated. (0): undigested DNA fragment; (M) is a pBR322 *Msp*I digest. Band sizes represent the distance between the restriction site and the labelled boundaries of the core particles. The orientations were determined by comparing the boundaries from successive restriction digestions. The X-shaped pattern is created by the shifting location of successive restriction sites relative to each core-particle boundary. (B) Diagram showing the various positions adopted by the histone octamer on the 172 bp repeating unit, as deduced from the data (31).

enzymes are required to unambiguously distinguish each of a pair of boundaries as upstream or downstream. An example of the application of this technique and its interpretation is shown in *Figure 3*.

The resolution of this technique can be as high as a few base pairs if denaturing polyacrylamide gels are employed. In non-denaturing gels, migration anomalies caused by restriction site overhangs and DNA bending can occur. The analysis benefits from using radioactively end-labelled, core-particle DNA, to yield pairs of restriction fragments with intensities reflective of position abundance. Thus, the frequency with which alternative nucleosome positioning sites become occupied during reconstitution can be assessed (16, 20, 31), as an indication of their relative affinity for the histone octamer.

5.2.3 Mapping by primer-extension

Synthetic, end-labelled oligonucleotides, complementary to the region being mapped, can be used to determine core-particle boundaries (32, 33). The oligonucleotide is hybridized in solution to chromatin-derived DNA fragments and extended with DNA polymerase or *Taq* polymerase. Extension terminates at the ends of the nucleosomal fragments such that the lengths of the products define the upstream boundaries with respect to the 5′ end of the oligonucleotide. An oligonucleotide extending towards the other end of the fragment or region would be used to identify the downstream boundaries of the same nucleosomes. This approach can be used to analyse core-particle DNA derived from reconstitutes of short fragments or plasmids and should provide quantitative, high-resolution mapping over short regions (≈ 250 bp). Any region of DNA can be assessed by this method by use of the appropriate oligonucleotide.

This approach has been used widely to analyse nucleosome arrays, particularly in yeast chromatin (32). For mapping nucleosome positioning in higher eukaryotic chromatin, a procedure incorporating ligation-mediated PCR (LMPCR) has been developed (Chapter 4 and ref. 34).

5.2.4 Mapping with mung bean nuclease

Core-particle DNA can be hybridized in solution to a long, single-stranded, 5′ end-labelled probe and digested with the single-stranded DNA-specific mung bean nuclease (16). The length of the nuclease-resistant, double-stranded products enables one boundary of positioned nucleosomes to be deduced. In a complementary experiment, a probe extending towards the other end of the fragment or region is used to identify the other boundary of the same nucleosomes. This approach can be used to analyse core-particle DNA derived from reconstitutes of short fragments or plasmids and provides quantitative, high-resolution mapping over short regions (≈ 250 bp).

5.2.5 Mapping by monomer extension

Core-particle DNA can itself be employed as a primer set to map nucleosome positions, using an approach termed monomer extension (16). This technique has particular advantages over some of the methods described above: extensive stretches (~ 1 kb) of reconstituted chromatin can be mapped, at high-resolution, in a single reaction, and quantitative information is provided concerning the relative affinity of the histone octamer for all the different positioning sequences. The procedure is outlined in *Figure 4*. The principal steps of the procedure are:

(a) Clone the DNA to be mapped in the same orientation into pBluescript (Stratagene) KS+ and KS– (or in opposing orientations in one vector), as this will facilitate independent mapping of both the upstream and downstream boundaries of the core particles (*Figure 4*). It is important to retain polylinker restriction sites which are unique to the whole construct as these are used in the monomer extension reactions.

(b) Transform the constructs into a host strain suitable for the preparation of ssDNA. We use *E. coli* strain DH11S (35). Prepare the ssDNA as

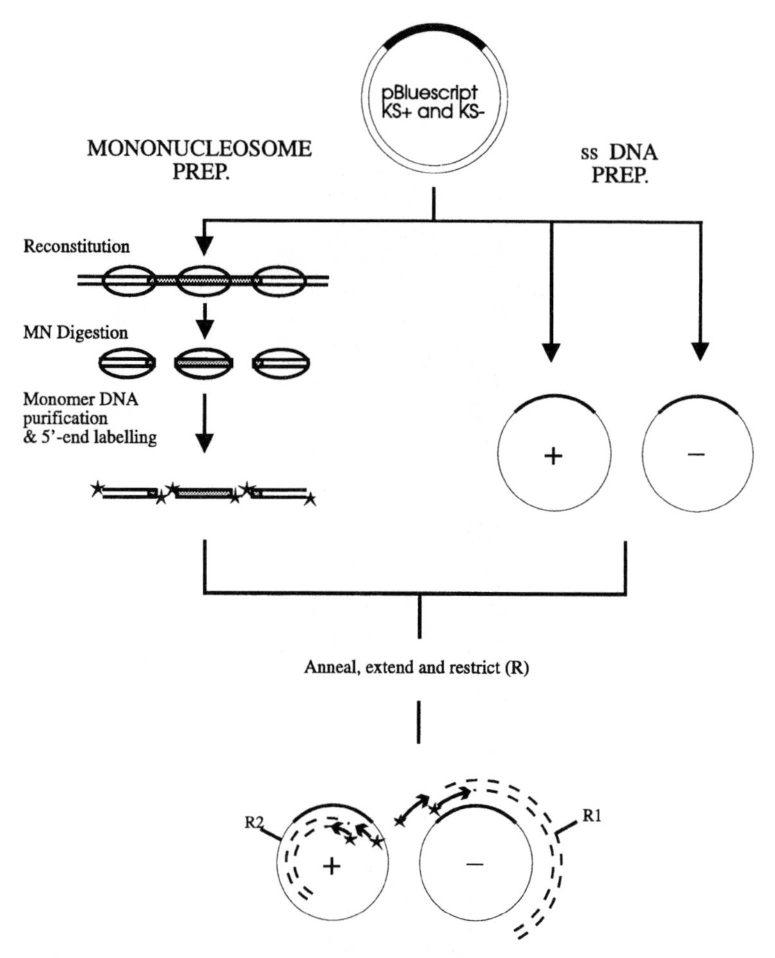

Figure 4. Schematic outline of the monomer extension procedure. Core-particle DNA is prepared from nucleosomes assembled on a plasmid containing the sequence to be mapped. These monomers are 5′end-labelled, denatured, and annealed to a single-stranded (ss) form of the plasmid. They are then extended by DNA polymerase in the presence of a restriction enzyme (R1) which cuts the nascent double-stranded DNA at a unique location just downstream of the region being mapped. Sizing the labelled DNA fragments produced locates the 5′-end of each priming core-particle DNA relative to this restriction site, thereby mapping one boundary of each histone octamer positioning site in the region of interest. The other boundary of each position can be mapped by extending the same monomers on the other strand of the plasmid.

described in *Protocol 5*. The quality of this template preparation is a principal determinant of the clarity of the monomer extension result.

(c) Reconstitute core histones (*Protocol 1*) onto one construct from the KS+, KS– pair and prepare core-particle DNA (*Protocol 3*). Label (36) and denature (37) the monomer DNA as described in *Protocol 6*.

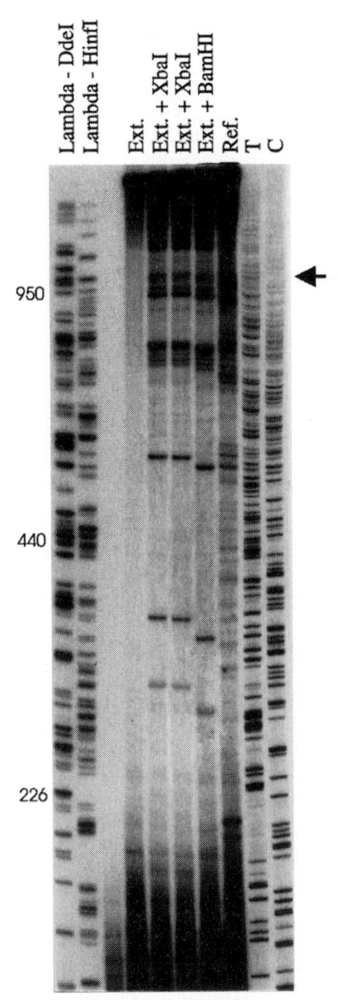

Figure 5. Polyacrylamide gel analysis of monomer extension products obtained from a 940 bp region of chicken β-globin sequence inserted into pBluescript KS–. Five identical annealing reactions were set up for extension in the absence (lane Ext.) or presence of different restriction enzymes: *Xba*I (in duplicate) corresponds to R1 in *Figure 4*; *Bam*HI cuts just 12 bp upstream from *Xba*I and gives an appropriately shifted version of the identical pattern; an enzyme cutting at the inappropriate end of the insert (R2 in *Figure 4*) reveals an entirely different pattern of positioning sites—from within the vector (lane Ref.). The large arrow indicates the limit of extension products obtained from the globin sequence; larger products map to the vector. The intense material at the bottom of all lanes derives from monomer self-priming.

(d) Anneal the monomer DNA to the single-stranded template and carry out the monomer extension reaction as described in *Protocol 7*. Reactions undertaken with a given template differ only in the restriction enzyme added during the extension phase and should comprise a minimum of two reactions: extension in the presence and absence of one restriction enzyme. A more comprehensive set of five reactions has been included in the example gel analysis illustrated in *Figure 5*.

(e) Size the extension products to map the boundaries of the positioning sites on the insert sequence relative to the restriction site employed in the monomer extension reaction.

Protocol 5. Preparation of the ssDNA template

Equipment and reagents

- TBG: 1.2% (w/v) bactotryptone (Difco), 2.4% (w/v) yeast extract (Difco), 0.4% (v/v) glycerol, 17 mM KH_2PO_4, 55 mM K_2HPO_4, 20 mM glucose added separately after autoclaving[a]
- M13KO7 (Pharmacia) helper phage ($\approx 10^{10}$ p.f.u./ml) prepared from a single plaque[b]
- 20% glycerol stock of the plasmid strain[a]
- 40% PEG/NaCl: 40% (w/v) PEG 6000, 2.5 M NaCl
- Buffered phenol, pH 8.0
- 20% PEG/NaCl: 20% (w/v) PEG 6000, 2.5 M NaCl
- TE buffer: 10 mM Tris–HCl, 0.1 mM EDTA, pH 8.0
- Chloroform:IAA (24:1)
- Buffered phenol:chloroform:IAA (25:24:1)
- Ampicillin and kanamycin
- 3 M sodium acetate
- SDS
- EDTA
- Beckman J2-21 centrifuge
- Beckman JA-14 rotor

Method

1. Inoculate 50 ml TBG with 200 μl of the glycerol stock of the plasmid strain and mix thoroughly. Add 500 μl of the M13KO7 stock. Incubate at 37 °C, 275 r.p.m. for 90 min.
2. Add ampicillin to 100 μg/ml and kanamycin to 70 μg/ml and incubate overnight.
3. Centrifuge to remove bacteria (7500 *g*, 20 min, 4 °C) and transfer the supernatant. Repeat.
4. Add 0.25 vol. of 40% PEG/NaCl and incubate on ice for 60 min[c].
5. Collect the phage particle precipitate (7500 *g*, 30 min), remove all residual PEG/NaCl by pipette and resuspend the pellet in 1.35 ml TE buffer. Add sodium acetate to 0.3 M[d] (150 μl 3 M, pH 5.2) and transfer to microcentrifuge tubes.
6. Extract with buffered phenol, then phenol:chloroform:IAA and chloroform:IAA.
7. Add 2.5 vol. of ethanol and incubate at – 70 °C for 60 min.
8. Collect the DNA by centrifugation (15 000 *g*, 15 min). Decant the supernatant and wash each pellet with 1 ml ice-cold 70% ethanol. Centrifuge for 5 min, decant the wash and dry the DNA.
9. Resuspend in a total of 600 μl TE buffer. Add SDS to 0.5% and EDTA to 3 mM and incubate at 56 °C for 30 min[e].
10. Add sodium acetate to 0.3 M and repeat steps 6 to 8. Resuspend in a total of 500 μl TE buffer.

11. Add 0.6 vol. (300 μl) 20% PEG/NaCl and incubate on ice for 60 min. Repeat step 8.
12. Resuspend the ssDNA in 200 μl TE buffer, such that the 40–50 μg typically obtained is at ≈ 200 ng/μl.

[a] See ref. 35.
[b] See ref. 38.
[c] The cloudiness is indicative of the yield of phage particles.
[d] A DNA precipitation concentration of sodium acetate, added here to raise the salt concentration and thereby aid dissociation of phage DNA from proteins which can then be more effectively removed during the phenol extractions. The pH of the sodium acetate is, in fact, unimportant because of buffering to ≈ pH 8.0 during the extractions; it applies equally in step 10.
[e] The conditions appear to benefit the removal of phage proteins.

Protocol 6. Labelling and denaturating the core-particle DNA

Equipment and reagents

- 10 × labelling buffer: 0.5 M Tris–HCl, pH 7.6, 0.1 M $MgCl_2$, 50 mM DTT, 1 mM spermidine, 1 mM EDTA[a]
- 10 units/μl T4 polynucleotide kinase (Amersham)
- [γ-^{32}P]ATP (3000 Ci/mmol; 10 m Ci/ml; Amersham)
- TE buffer (*Protocol 5*)
- Phenol:chloroform:IAA
- Chloroform:IAA
- 5 M ammonium acetate, pH 7.5
- Sephadex G-50 (Pharmacia)
- Low-speed, bench-top centrifuge with swing-out rotor (for step 3)

Method

1. Combine 300 ng[b] of monomer DNA with water to a total volume of 8.5 μl. Add 1.5 μl of 10 × labelling buffer, 4 μl of [γ-^{32}P]ATP and 1 μl (10 units) T4 polynucleotide kinase. Incubate at 37 °C for 45 min[c].
2. Add 30 μl TE buffer and extract with phenol:chloroform:IAA then chloroform:IAA (*Protocol 5*).
3. Separate the DNA from unincorporated label by spun column chromatography[d] using a 1 ml disposable syringe packed with Sephadex G-50 equilibrated into water.
4. To the 45 μl of monomers eluted in water, add 5 μl 2 M NaOH and mix thoroughly. Incubate at room temperature for 10 min.
5. Add 20 μl 5 M ammonium acetate (pH 7.5) and mix thoroughly. Add 30 μl water and mix thoroughly. Add 250 μl ethanol and incubate at −70 °C overnight.
6. Recover and wash the DNA as described in *Protocol 5*, step 8 and resuspend in 30 μl TE buffer (10 ng DNA/μl)[e].

[a] See ref. 36.
[b] This is sufficient for up to 10 monomer extension reactions at 30 ng DNA per reaction.
[c] During this reaction, the 3′ phosphatase activity of the kinase generates the 3′ hydroxyl ends essential to the use of monomers as primers.
[d] See ref. 38.
[e] Proceed directly to *Protocol 7* as fresh monomers give the best results.

Protocol 7. Monomer extension reaction

Equipment and reagents

- ssDNA template
- Labelled, denatured monomer DNA (*Protocol 6*)
- Pre-mix 1: 2.5 μl 1 M NaCl, 2.5 μl 20 mM DTT, 2 μl 250 mM Tris–HCl, pH 7.5, 2 μl 250 mM $MgCl_2$, 8 μl water
- dNTPs: dATP, dCTP, dGTP, dTTP (Pharmacia), each at 250 μM
- Pre-mix 2: 18 μl water, 2.5 μl 2 mg/ml BSA, 2 μl 250 μM dNTPs
- Appropriate restriction enzymes
- 10 units/μl Klenow DNA polymerase (Pharmacia)
- Phenol:chloroform:IAA
- Chloroform:IAA
- sodium acetate
- Denaturing 6% polyacrylamide gel
- Thermal cycler
- Sequencing type electrophoresis apparatus (e.g. Sequi-gen, Bio-Rad)
- Denaturing gel loading buffer; 90% (v/v) formamide, 1 × TBE, 10 mM EDTA, 0.05% bromophenol blue, 0.05% xylene cyanol.

Method

1. For each reaction (25 μl)[a], mix 1 μg ssDNA template (5 μl) with 30 ng (3 μl) labelled, denatured monomer DNA[b]. Add 17 μl of pre-mix 1 (final concentrations: 100 mM NaCl, 2 mM DTT, 20 mM Tris, 20 mM $MgCl_2$).
2. Denature at 95 °C for 3 min, hold at 80 °C for 30 sec, then cool slowly to 55 °C over ≈ 45 min.
3. To each annealing reaction add 22.5 μl of pre-mix 2.
4. Add 20 units (2 μl) of the chosen restriction enzyme (or 2 μl water for the control) then 5 units (0.5 μl) of Klenow DNA polymerase and mix thoroughly by pipette. Incubate at 37 °C for 60 min.
5. Purify by extraction with phenol:chloroform:IAA and chloroform:IAA (*Protocol 5*).
6. Add sodium acetate to 0.3 M (*Protocol 5*) and precipitate the DNA –(*Protocol 5*, steps 7 and 8). Resuspend each DNA pellet in 5–10 μl denaturing gel loading buffer.
7. Heat-denature the samples and the size standards[c] and electrophorese in a denaturing 6% polyacrylamide gel. Retain fragments of about 175 nt and larger, running the unextended monomer DNA off the gel.

[a] If only two reactions (i.e. plus and minus restriction enzyme) are to be undertaken, then set up a double volume annealing (50 μl) and maintain as a single reaction until step 4, then split it in half for addition of water or restriction enzyme.

[b] Aim to use about a 2.5-fold molar excess of ss template over monomer primer strand: for a 4 kb (pBluescript with 1 kb insert) ss template, 1 μg equates to ~ 0.75 pmoles; 30 ng of 146 bp monomer supplies 15 ng of the strand extendible on this particular strand of the template and equates to ≈ 0.3 pmoles.

[c] Sequencing reactions and restriction fragments (up to ≈ 1 kb) provide the best combination of size standards: we use ^{35}S-labelled C and T sequencing reactions of M13mp18 ssDNA (Pharmacia) using the – 20 Universal forward primer and ^{32}P end-labelled *Hin*fI and *Dde*I digests of phage lambda.

Monomer extension is limited to mapping regions of ≤ 1 kb because of the inability to size larger extension products with any accuracy. In order to map positioning sites over a contiguous region of several kb, clone the region into pBluescript in its entirety and as a series of overlapping ~1 kb fragments. Generate monomers from the construct containing the complete region and extend, in turn, on ssDNA generated from each of the subclones. A 2–300 bp overlap between adjacent subclones will negate the obscuring effect on gels of monomer self-priming (see *Figure 5*). We have used this strategy to produce the first long-range, high-resolution nucleosome positioning map for an entire, contiguous gene region (39).

Subcloning can be avoided on the rare occasion of unique restriction sites being appropriately spaced throughout the region of interest. Fortunately, one can create this situation for any DNA of known sequence. In this approach, a single ssDNA template is generated from the construct containing the entire region of interest. This template is then linearized just downstream of any section to be mapped by annealing an oligonucleotide complementary to a chosen restriction site and its flanking sequence (6 bp recognition site and 5 or 6 bp either side): this creates a short double-stranded region which is uniquely cut upon addition of the restriction enzyme. Monomers are then annealed to this pre-linearized template and extended (in the absence of a restriction enzyme). By designing a series of oligonucleotides which anneal, and thereby enable linearization of the template, every 500–1000 bp, one can 'walk' the mapping along a given strand of the insert.

Acknowledgements

The authors acknowledge the support of the Wellcome Trust. We thank Ali Yenidunya and Geert Meersseman for their substantial contribution in developing some of the techniques described here.

References

1. Travers, A. A. (1987). *TIBS*, **12**, 108.
2. Simpson, R. T. (1991). *Prog. Nucl. Acid Res. Mol. Biol.*, **40**, 143.
3. Wolffe, A. P. (1994). *TIBS*, **19**, 240.
4. Drew, H. R. and Travers, A. A. (1985). *J. Mol. Biol.*, **186**, 773.
5. Satchwell, S. C., Drew, H. R., and Travers, A. A. (1986). *J. Mol. Biol.*, **191**, 659.
6. Calladine, C. R. and Drew, H. R. (1986). *J. Mol. Biol.*, **192**, 907.
7. Richmond, T. J., Finch, J. T., Rushton, B., Rhodes D., and Klug, A. (1984). *Nature* (London), **311**, 532.
8. Hayes, J. J., Tullius, T. D., and Wolfe, A. P. (1990). *Proc. Natl Acad. Sci. USA*, **87**, 7405.
9. Buckle, R., Balmer, M., Yenidunya, A., and Allan, J. (1991). *Nucl. Acids. Res.*, **19**, 1219.
10. Rhodes, D. and Laskey, R. A. (1989). In *Methods in enzymology* (eds. Wassarman P. M. and Kornberg, R. D.), Academic Press, San Diego, California. Vol. 170, p. 575.

11. Stein, A. (1989). In *Methods in enzymology* (eds. Wassarman P. M. and Kornberg, R. D.), Academic Press, San Diego, California. Vol. 170, p. 585.
12. Lutter, L. C. (1989). In *Methods in enzymology* (eds. Wassarman P. M. and Kornberg, R. D.), Academic Press, San Diego, California. Vol. 170, p. 264.
13. Ramsay, N., Felsenfeld, G., Rushton, B. M., and McGhee, J. D. (1984). *EMBO J.*, **3**, 2605.
14. Shrader, T. E. and Crothers, D. M. (1989). *Proc. Natl Acad. Sci. USA*, **86**, 7418.
15. Kefalas, P., Gray, P., and Allan, J. (1988). *Nucl. Acids Res.*, **16**, 501.
16. Yenidunya, A., Davey, C., Clark, D., Felsenfeld, G., and Allan, J. (1994). *J. Mol. Biol.*, **237**, 401.
17. Drew, H. R. and Calladine, C. R. (1987). *J. Mol. Biol.*, **195**, 143.
18. Germond, J. E., Bellard, M., Oudet, P., and Chambon, P. (1976). *Nucl. Acids Res.*, **3**, 3173.
19. Rhodes, D. (1985). *EMBO J.*, **4**, 3473.
20. Zhang, X-Y., Fittler, F., and Hörz, W. (1983). *Nucl. Acids Res.*, **11**, 4287.
21. Meersseman, G., Pennings, S., and Bradbury, E. M. (1992). *EMBO J.*, **11**, 2951.
22. Pennings, S., Meersseman, G., and Bradbury, E. M. (1994). *Proc. Natl Acad. Sci. USA.*, **91**, 10275.
23. Crothers, D. M., Gartenberg, M. R., and Shrader, T. E. (1991). In *Methods in enzymology* (ed. Sauer, R. T.). Academic Press, San Diego, California. Vol. 208, p. 118.
24. Pennings, S., Meersseman, G., and Bradbury, E. M. (1991). *J. Mol. Biol.*, **220**, 101.
25. Duband-Goulet, I., Carot, V., Douc-Rasy, S., and Prunell, A. (1992). *J. Mol. Biol.*, **224**, 981.
26. Pennings, S., Meersseman, G., and Bradbury, E. M. (1992). *Nucl. Acids Res.*, **20**, 6667.
27. Tullius, T. D., Dombroski, B. A., Churchill, M. E. A., and Kam, L. (1987). In *Methods in enzymology* (ed. Wu, R.). Academic Press, San Diego, California. Vol. 155, p. 537.
28. Linxweiler, W. and Hörz, W. (1986). *Cell*, **42**, 281.
29. Morse, R. H. (1989). *EMBO J.*, **8**, 2343.
30. Almer, A., Rudolph, H., Hinnen, A., and Hörz, W. (1986). *EMBO J.*, **5**, 2689.
31. Meersseman, G., Pennings, S., and Bradbury, E. M. (1991). *J. Mol. Biol.*, **220**, 89.
32. Shimizu, M., Roth, S. Y., Szent-Gyorgyi, C., and Simpson, R. T. (1991). *EMBO J.*, **10**, 3033.
33. Bresnsick, E. H., Rories, C., and Hager, G. L. (1992). *Nucl. Acids Res.*, **20**, 865.
34. Pfeifer, G. P. and Riggs, A. D. (1991). *Genes Dev.*, **5**, 1102.
35. Lin, J-J., Smith, M., Jessee, J., and Bloom, F. (1992). *BioTechniques*, **12**, 718.
36. Maniatis, T., Fritsch, E. F., and Sambrook, J. (1982). *Molecular cloning, a laboratory manual* (first edition). Cold Spring Harbor Laboratory Press, Cold Spring Harbor, NY.
37. Hattori, M. and Sakaki, Y. (1986). *Anal. Biochem.*, **152**, 232.
38. Sambrook, J., Fritsch, E. F., and Maniatis, T. (1989). *Molecular cloning, a laboratory manual* (second edition). Cold Spring Harbor Laboratory Press, Cold Spring Harbor, NY.
39. Davey, C., Pennings, S., Meersseman, G., Wess, T. J., and Allan, J. (1995). *Proc. Natl Acad. Sci. USA.*, **92**, 11210.

9

Transcriptional and structural analysis of chromatin assembled *in vitro*

MICHAEL J. PAZIN and JAMES T. KADONAGA

1. Introduction

The biochemical analysis of the transcriptional properties of DNA packaged into chromatin, as it is *in vivo*, promises to increase our understanding of gene regulation (1–4). This chapter focuses on the assembly and transcription of chromatin by using two different extracts derived from *Drosophila melanogaster* embryos. We have also included techniques for the analysis of chromatin structure to verify the proper assembly of chromatin as well as to examine mechanisms of gene regulation. Hence, this chapter describes methods for the assembly of chromatin (*Protocols 1–3*), *in vitro* transcription of the reconstituted chromatin templates (*Protocol 4*), and analysis of chromatin structure and the binding of transcription factors to the nucleosomal templates (*Protocols 5–7*).

2. General considerations

2.1 Reconstitution of chromatin: purified systems vs. crude extracts

Chromatin can be reconstituted with either purified components or crude extracts. The existing purified reconstitution systems have the advantage of producing templates of defined composition, but typically do not yield chromatin with regularly spaced nucleosomes and a physiological nucleosome repeat length. On the other hand, crude extracts derived from either *Xenopus laevis* or *Drosophila* can efficiently assemble chromatin with properties that are similar to those of native chromatin, but the composition of the resulting chromatin is less well defined. The assembly factors in these extracts that are responsible for the physiological nucleosome spacing have yet to be purified and characterized. Thus, the use of either the purified or the crude chromatin reconstitution systems will be dependent upon the nature of the experiment.

For instance, the purified reconstitution systems (such as salt dilution or dialysis) are generally used for the preparation of mononucleosomes, whereas the crude extracts are often preferred for the assembly of regularly spaced nucleosomal arrays. In the following four subsections, we will briefly describe some commonly used methods for the reconstitution of chromatin.

2.1.1 Salt dialysis/dilution

At a high salt concentration (such as 2.0 M NaCl, a concentration at which core-histone octamers will dissociate from the DNA), the DNA template is mixed with either purified core histones or bulk native chromatin (from which core histones will be released at 2.0 M NaCl), and the salt concentration is subsequently reduced by dialysis or dilution, which results in the reconstitution of nucleosome cores onto the template (see ref. 5 for a review). This technique has been widely used for the reconstitution of mononucleosomal cores on short DNA fragments (150–200 bp), but typically does not assemble physiological nucleosomal arrays on long DNA fragments, which is a limitation for functional studies, such as transcriptional analyses.

2.1.2 Polyglutamic acid

Polyglutamate, a polyanion, can be used to facilitate the deposition of nucleosomes (for a review, see ref. 6). Core histones are combined with high molecular weight polyglutamate to form core histone–polyglutamate complexes. These complexes are then added to DNA, and the core histones are transferred from the polyglutamate to the DNA to yield nucleosomes. The efficiency of nucleosome deposition by this method can vary widely with different preparations of polyglutamate. In addition, chromatin prepared with polyglutamate does not usually possess regular nucleosomal arrays.

2.1.3 *Xenopus* extracts

Xenopus eggs and oocytes contain large quantities of histones stored in their cytoplasm as complexes with histone binding proteins, which include nucleoplasmin and N1/N2. These complexes provide the histones necessary for chromatin assembly during the early embryonic nuclear divisions, which occur in the absence of protein synthesis. Crude S-150 extracts prepared from *Xenopus* oocytes can efficiently assemble regularly spaced nucleosomal arrays in the presence of an ATP regenerating system and DNA (Chapter 10 and ref. 7). These oocyte extracts presumably use the core histone-containing complexes, perhaps along with other factors, for the assembly of chromatin with physiological nucleosome spacing.

Heat-treated *Xenopus* egg extracts, which contain nucleoplasmin and N1/N2, are also commonly used for the reconstitution of chromatin (for a review, see ref. 8). These heat-treated extracts do not, however, yield arrays of nucleosomes with physiological spacing, possibly because of heat denaturation of chromatin assembly activities.

2.1.4 *Drosophila* extracts

Chromatin with physiological nucleosome spacing can be efficiently assembled with extracts derived from *Drosophila* embryos (refs 9–11), which are similar to the *Xenopus* S-150 extract described above. Furthermore, large quantities (hundreds of grams) of embryos can be obtained at low cost, without the seasonal variation in extract quality that is commonly observed with *Xenopus* eggs and oocytes. Extracts prepared from predominantly post-blastoderm embryos (10, 11) assemble chromatin with excellent transcription capacity (10, 12, 13).

In this chapter, we will describe the use of the S-190 extract derived from *Drosophila* embryos (10, 11) for the assembly of chromatin that is to be subsequently subjected to transcriptional and structural analyses. The choice of this extract is based, in part, on its ability to generate regularly spaced nucleosomal arrays with physiological nucleosome spacing in an efficient and reproducible manner.

2.2 Histone H1

In metazoans, native chromatin is estimated to contain, on average, approximately one molecule of the linker histone H1 (or the H1 variant, H5) per nucleosomal core (14), while, in contrast, the yeast *Saccharomyces cerevisiae* does not appear to have a homologue of histone H1 (see discussion in ref. 1). The available data suggest that histone H1 contributes to the repression of transcription (1–4). If desired, the role of histone H1 in transcriptional regulation can be tested biochemically by comparing the properties of chromatin templates assembled in the presence or absence of exogenously added H1 (see, for example, refs 10, 12).

2.3 Transcription of pre-assembled chromatin templates

In vivo, some genes can be activated in the absence of DNA replication, while others appear to require replication, which might facilitate binding of transcription factors to the DNA template by disrupting repressed chromatin structure (1–4). Similarly, biochemical studies have revealed that some transcription factors can activate transcription when added to fully assembled chromatin templates (10, 12, 13), whereas other factors can activate transcription only if pre-bound to naked DNA prior to chromatin assembly (15). Thus, it may be important to consider whether or not a transcription factor (or a combination of factors) is able to activate transcription from a pre-assembled chromatin template.

2.4 Non-specific inhibition of chromatin assembly

In studies that involve the characterization of chromatin templates, it is important to note that any factor that contains either an inhibitor of chromatin assembly or an agent that disrupts chromatin structure will relieve the chromatin-mediated repression of transcription and artefactually appear to activate transcription. It is therefore essential to examine the structure of the chromatin templates that are assembled in the presence versus the absence of putative activators. Moreover, the transcriptional effects of a sequence-specific DNA binding factor can be tested with parallel DNA templates that either do or do not contain the recognition site for the factor. In this manner, the transcriptional effects can be determined to be dependent upon both the factor binding site as well as the factor itself (12).

2.5 Simultaneous structural and functional analyses

With the protocols described in this chapter, each preparation of newly assembled chromatin is simultaneously subjected to both transcriptional and structural analyses. In this manner, the structure of the identical sample of chromatin that was transcribed is known. For convenience, stopping points are indicated in the protocols, so that the analyses may be finished separately, if desired.

3. Chromatin assembly

3.1 Assembling the template DNA into chromatin

First, a chromatin assembly system must be chosen by considering some of the issues that were discussed earlier (as mentioned above, we will describe the use of the *Drosophila* S-190 extract). It must also be decided what accessory factors (e.g. transcriptional activators and/or histone H1) will be added to the reaction as well as the timing of the addition of these factors (i.e. before, during, or after assembly). Sufficient chromatin must be assembled to perform the transcription reactions, controls for the efficiency of assembly, and perhaps other structural analyses. *Protocol 1* describes a sample assembly reaction, and indicates some of the variable parameters. We have assembled chromatin in 125 ng to 5 μg amounts by proportionate scaling of the reaction presented here. Additional details of assembly reactions are described in ref. 11. For some experiments, it may be necessary to purify the chromatin by sucrose gradient sedimentation and/or size exclusion chromatography (11). The chromatin should be used immediately after assembly.

3.2 Is the reconstituted chromatin good enough for transcriptional studies?

The quality of the chromatin template is critical to the success of any transcription experiment. In general, it is probably best to reconstitute chromatin that resembles native chromatin as closely as possible. Hence, in the reconstitution of a yeast DNA gene into chromatin, histone H1 would be omitted, whereas in studies of *Drosophila* or human genes, it would be more appropriate to assemble chromatin containing about one molecule of H1 per nucleosome. In addition, chromatin should be assembled with high efficiency, because transcription from incompletely assembled chromatin could be predominantly from nucleosome-free regions.

Protocol 1. Chromatin assembly—1 μg reaction

Reagents

- Buffer R: 10 mM Hepes, K^+, pH 7.5, containing 10 mM KCl, 1.5 mM $MgCl_2$, 0.5 mM EGTA, 10% (v/v) glycerol, 10 mM β-glycerophosphate (Sigma, G-6251), 1 mM dithiothreitol (DTT; US Biochemicals, 15397), 0.2 mM phenylmethylsulfonyl fluoride (PMSF; Sigma, P-7626). Store in 10 ml aliquots at −20°C. (This can be freeze–thawed many times.)
- S-190 chromatin assembly extract[a]
- ~1 mg/ml purified *Drosophila* core histones[b]
- ~2 mg/ml purified *Drosophila* histone H1 (if desired)[c]
- Creatine phosphate: 0.5 M phosphocreatine (Sigma, C-3755) dissolved in 20 mM Hepes, K^+, pH 7.6, and adjusted to pH 7 with NaOH. Store in 1 ml aliquots at −20°C. (Can freeze–thaw many times.)
- Transcriptional activator (if desired)
- ATP: 0.5 M adenosine triphosphate, adjusted to pH 7 with NaOH. Store at −20°C.
- Creatine kinase: 5 mg/ml creatine phosphokinase (Sigma C-3755) in 10 mM potassium phosphate pH 7, containing 50 mM NaCl and 50% (v/v) glycerol. Store in 5 μl aliquots at −80°C. (Use once, then discard.)
- Mg–ATP mix. For 100 μl Mg–ATP mix, combine: 22.7 μl water, 42.5 μl 0.5 M creatine phosphate, 4.25 μl 0.5 M ATP, 0.84 μl 5 mg/ml creatine kinase, and 29.7 μl 0.1 M $MgCl_2$ at room temperature. (Make immediately before adding to the reconstitution reactions.)
- ~1 mg/ml plasmid DNA template in TE buffer. Prepare by two successive CsCl equilibrium gradient purification steps.
- TE buffer: 10 mM Tris–HCl, pH 8, 1 mM EDTA

Method

1. Combine (in order, flicking the tube gently after each addition) buffer R, S-190 (50 μl), core histones (0.96 μg), histone H1 (0.9 μg). Adjust the volume of buffer R so that the final volume after this step is 170 μl. Incubate for 30 min at room temperature.
2. Add (in order, flicking the tube gently after each addition) Mg–ATP mix (28.5 μl), activator (optional), and DNA (1 μg) to a final volume of approximately 200 μl.
3. Incubate for 5 h at 27 °C. (If desired, add additional factors 4.5 h into this step, when assembly is complete. Reconstitution is incomplete until after 4 h. If the incubation proceeds for longer than 6 h, however, an endogenous nuclease will begin to degrade the DNA.)
4. Use the chromatin immediately for transcriptional and structural analyses.

[a] Prepare as described in ref. 11. Store in 1 ml aliquots at −80°C. It can be freeze–thawed at least twice.
[b] Purify as described in ref. 11. Store in aliquots at − 80 °C. This can be freeze–thawed many times.
[c] Purified as described in ref. 16. Store in aliquots at −80 °C. Can freeze–thaw many times.

The efficiency of nucleosome deposition on circular DNA templates can be estimated by DNA supercoiling analysis (see Chapter 7 and refs 10 and 17). In the presence of topoisomerase I (which is present in the S-190 extract), the deposition of one nucleosome causes a change in the linking number of –1, and thus, the number of nucleosomes on a template can be measured by the change in the linking number of the deproteinized chromatin relative to a relaxed DNA standard by carrying out one- or two-dimensional, agarose gel electrophoresis under conditions that resolve individual topoisomers (Chapter 7). In the interpretation of this assay, however, it is important to note that non-histone proteins as well as subnucleosomal particles, such as H3–H4 tetramers, will also supercoil DNA (18). Hence, it is important to use the supercoiling assay in conjunction with other analyses, such as the micrococcal digestion assay (described below), for the characterization of reconstituted chromatin.

The DNA supercoiling assay can also be used to determine the relative efficiency of nucleosome deposition in a series of assembly reactions by

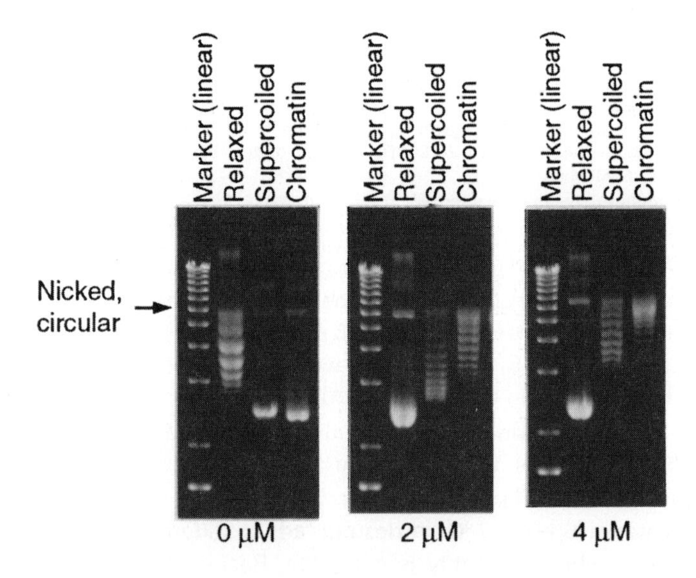

Figure 1. Chromatin was assembled as in *Protocol 1*, and the topoisomers were characterized on one-dimensional supercoiling gels. After chromatin assembly, the DNA was deproteinized and analysed as in *Protocol 2*. For comparison, the same 3.2 kb plasmid (pGIE-0) is shown as the starting material (highly supercoiled by the bacterial host) and after being relaxed with topoisomerase I. The same samples were run on gels containing different chloroquine concentrations to demonstrate the effect of the chloroquine. In some of the lanes, nicked DNA is visible at the position indicated by the arrow. The marker is the 1 kb ladder from Gibco.

subjecting the deproteinized chromatin samples to one-dimensional, agarose gel electrophoresis in the presence of chloroquine (a DNA intercalating agent that introduces positive supercoiling into the DNA and allows better resolution of highly negatively supercoiled DNA in a standard, one-dimensional agarose gel) (*Protocol 2*; see also *Protocol 3*, Chapter 7) (*Figure 1*). This assay can be used to determine whether an activator protein (or a contaminant in the protein fraction) inhibits the assembly of chromatin, which would be seen as a decrease in the degree of negative supercoiling. (Remember that an inhibitor of chromatin assembly might artefactually appear to be a transcriptional activator.) For comparison, we often analyse samples on two agarose gels—one with and one without chloroquine. In addition, it is important to note that the optimal concentration of chloroquine will vary according to the specific experimental conditions, which include the size of the plasmid and the DNA sequence of the plasmid. It is therefore necessary to determine empirically the optimal chloroquine concentration.

Protocol 2. One-dimensional supercoiling assay[a]

Equipment and reagents

- 40 μl chromatin (containing 200 ng DNA), freshly prepared (*Protocol 1*)
- 2.5 mg/ml proteinase K (USB, 20818) stock solution in TE buffer. Store at − 20 °C. (Can freeze–thaw many times.)
- Transcription stop solution: 20 mM Na-EDTA, pH 8, containing 0.2 M NaCl, 1% (w/v) SDS, 0.25 mg/ml glycogen (Sigma, G-0885). Store at room temperature.
- Transcription stop solution–proteinase K: combine transcription stop solution and proteinase K in a 20:1 (v/v) ratio before adding to samples. Make fresh each time.
- Phenol:chloroform:isoamyl alcohol (25:24:1 (v/v/v))
- 2.5 M ammonium acetate
- 1 × Tris/borate/EDTA (TBE) buffer: 0.09 M Tris–borate, 1 mM EDTA, pH 8.0
- 100 mM chloroquine diphosphate (Sigma, C-6628) in water. Store at 4 °C in the dark.
- RNase (DNase-free) (This is available as a solution containing 0.5 mg/ml protein; Boehringer Mannheim, 1119 915)
- 5 × TG loading buffer: 50% (v/v) glycerol, 5 mM EDTA, pH 8.0, 0.1% (w/v) Bromophenol blue
- 1% agarose gel
- 0.75 μg/ml ethidium bromide in water
- Rotary concentrator (e.g. Speed-Vac, Fisher Scientific)

Method

1. Pipette 40 μl of chromatin (containing 200 ng DNA) into a 1.5 ml microcentrifuge tube. (This sample can be stored at − 20 °C after this step.)
2. Add 40 μl of transcription stop solution–proteinase K to the chromatin. Incubate for 15 min at 37 °C.
3. Extract with phenol:chloroform:isoamyl alcohol (80 μl). Centrifuge for 5 min at 15 000 *g* in a microcentrifuge at room temperature.
4. Transfer the aqueous phase to a new tube containing 2.5 M ammonium acetate (8 μl) and ethanol (180 μl). Mix by vortexing.
5. Centrifuge for 15 min at 15 000 *g* at room temperature in a microcentrifuge.

Protocol 2. *Continued*

6. Remove the supernatant, and dry the pellet in a rotary concentrator (e.g. Speed-Vac).
7. Resuspend samples in 4 μl of 1 × TBE buffer containing 2 μM chloroquine diphosphate (freshly added) and 0.4% (v/v) RNase (DNase-free). Incubate for approximately 15 min at room temperature.
8. Add 1 μl of 5 × TG loading buffer.
9. Analyse samples on a 1% (w/v) agarose gel prepared with 1 × TBE containing 2 μM chloroquine diphosphate. Run the gel in the same buffer. Note that the chloroquine must be added when the agarose has cooled to approximately 55 °C. Do not add ethidium bromide to the gel or buffer—it will alter the supercoiling of the samples.
10. Stain gels with 0.75 μg/ml ethidium bromide in water for 15 min. Visualize DNA with a UV transilluminator.

[a] cf. *Protocol 3* in Chapter 7.

Bulk, native chromatin has nucleosomes in periodic (regularly spaced) arrays. The regularity of the spacing and the repeat length (distance from nucleosome to nucleosome, in bp) can be measured by micrococcal nuclease digestion analysis (Chapter 3 and ref. 19). In this assay, the chromatin is partially digested with micrococcal nuclease, and the resulting DNA fragments are deproteinized and resolved by agarose gel electrophoresis (*Protocol 3*). By using this technique, the regularity of nucleosomal spacing and nucleosome repeat length of the reconstituted chromatin can be compared with those of native chromatin.

Protocol 3. Micrococcal nuclease analysis of chromatin[a]

Equipment and reagents

- 200 μl chromatin (containing 1 μg DNA), freshly prepared (*Protocol 1*)
- 200 U/ml micrococcal nuclease (MNase) stock solution. Dissolve 200 units micrococcal nuclease (Sigma, N-5386) in 1 ml of 5 mM sodium phosphate buffer, pH 7.0, containing 2.5 μM $CaCl_2$. Store in aliquots at –20 °C. (Can freeze–thaw many times.)
- Buffer R (*Protocol 1*)
- Micrococcal nuclease dilutions. For each sample of chromatin, prepare the following dilutions of the micrococcal nuclease stock in buffer R (make immediately before use): 1.3 U/ml; 4 U/ml; 12 U/ml; 36 U/ml.
- RNase (DNase-free) (*Protocol 2*)
- 2.5 M ammonium acetate
- MNase stop solution: 2.33 μl TE buffer (*Protocol 1*), 2.38 μl 0.5 M EDTA, and 0.29 μl RNase (DNase-free) per sample. Make up a cocktail for the desired number of samples.
- Transcription stop solution–proteinase K (*Protocol 2*)
- Phenol:chloroform:isoamyl alcohol (*Protocol 2*)
- Rotary concentrator (e.g. Speed-Vac)
- 5 × TG loading buffer (*Protocol 2*)
- 20 × 20 cm 1.25 % (w/v) agarose gel
- Tris–glycine buffer: 0.22 M Tris base, 1.92 M glycine
- 0.75 μg/ml ethidium bromide in water

Method

1. Add 50 μl of chromatin (containing 250 ng of DNA) to a 1.5 ml microcentrifuge tube containing 0.1 M $CaCl_2$ (1.5 μl). Mix immediately by flicking the tube gently.
2. Add 5 μl of a micrococcal nuclease dilution. Mix immediately by flicking the tube gently. Start multiple samples consecutively, 15 sec apart. (In this protocol, we describe the use of several concentrations of MNase to achieve differing degrees of digestion. Alternatively, a constant concentration of MNase could be used, and the time of digestion could be varied (e.g. ref. 9).
3. Digest for 10 min at room temperature.
4. Add MNase stop solution (5 μl). Mix immediately by flicking the tube gently. Stop multiple samples consecutively, 15 sec apart. Incubate for approximately 15 min at room temperature. (Can store samples at −20 °C after this step.)
5. Add 100 μl of transcription stop solution–proteinase K. Vortex. Incubate for 15 min at 37 °C.
6. Extract with 150 μl phenol:chloroform:isoamyl alcohol. Centrifuge for 5 min at 15 000 *g* at room temperature in a microcentrifuge.
7. Transfer the aqueous phase to a new tube containing 2.5 M ammonium acetate (15 μl) and ethanol (340 μl). Vortex. Centrifuge for 15 min at 15 000 *g* at room temperature in a microcentrifuge. (Do not use sodium acetate, as it will give a large, oily pellet.)
8. Remove the supernatant, and dry the pellet in rotary concentrator (e.g. Speed-Vac).
9. Resuspend the pellet in 5 μl of TE buffer. Let it stand for 10 min at room temperature. Add 1.25 μl of 5 × TG loading buffer (*Protocol 2*).
10. Run on a 1.25 % (w/v) agarose gel in Tris–glycine buffer (20 × 20 cm, 150 ml) at 4 °C and 200 V. Stop the gel when the Bromophenol blue dye has migrated two-thirds of the length of the gel.
11. Stain the gel with 0.75 μg/ml ethidium bromide in water for 30 min at room temperature. Visualize DNA with a UV transilluminator. The gel can be destained in water for 2 h at room temperature to reduce the background fluorescence.

[a] cf. *Protocol 4*, Chapter 3.

When chromatin is assembled onto a small (e.g. 3.1 kbp) plasmid with the *Drosophila* S-190 extract, the micrococcal nuclease assay typically reveals about 8–12 nucleosomal bands with a physiological repeat length. In fact, all 16 nucleosomal bands are often observed with a 3.1 kb plasmid. (In *Drosophila*, the native repeat length is 190 bp (11, 20), and thus, a 3.1 kb minichromosome would be expected to have 16 nucleosomes.) The repeat length is dependent upon the nucleosome density, the incorporation of histone H1 (which increases the repeat length), and the completeness of assembly. The number of nucleosomal bands seen is dependent upon the regularity of the nucleosome spacing, the completeness of assembly, and the quality (e.g. presence of residual chemical contaminants, etc.) of the samples. An example of a micrococcal digestion ladder that was visualized by autoradiography rather than staining with ethidium bromide can be seen in *Figure 4A*. If the chromatin does not have at least five discrete bands in the micrococcal ladder, it should probably not be used for transcriptional studies.

Electron microscopy is not routinely used for the examination of reconstituted chromatin, but direct visualization of the reconstituted chromatin provides important evidence for the complete assembly of nucleosomal arrays (Chapter 6 and refs 10, 11). For instance, as noted above, non-histone chromosomal proteins, such as HMG proteins (see, for example, ref. 21), will also introduce negative supercoils into DNA, and may thus lead to misinterpretation of DNA supercoiling data. In addition, electron microscopy can reveal structural features of the reconstituted chromatin such as the zigzag nature of H1-containing chromatin as well as folding of the chromatin into the 30 nm diameter filament (Chapter 6).

4. Transcription

After determining the integrity of the reconstituted chromatin, the transcriptional analysis is straightforward. We directly transcribe chromatin templates under conditions that are identical to those used for naked DNA templates. We transcribe the chromatin at the same time as any nuclease digests, such as *Protocols 3* and 5. Typically, a histone-deficient *Drosophila* nuclear extract, termed the soluble nuclear fraction (SNF) (22, 23), is used as the source of basal transcription factors (10, 12). HeLa nuclear extracts can also be used (13).

The transcription reaction could be performed with or without assembly of the transcription pre-initiation complex prior to the addition of the four ribonucleoside 5′-triphosphates (rNTPs). If the SNF is pre-incubated with the templates prior to the addition of the rNTPs, then the pre-initiation complex will be assembled, wherefrom initiation and elongation would occur rapidly upon the subsequent addition of the rNTPs. On the other hand, it is also possible to add the SNF and the rNTPs simultaneously, wherein the transcription complexes would initiate immediately upon assembly. In our experience, identical results are obtained with either transcription protocol; the latter version, described in *Protocol 4*, is technically simpler. In some instances, however, it will be necessary to assemble the pre-initiation complexes prior to the addition of the rNTPs, as in the analysis of single-round transcription reactions. In such instances, *Protocol 4* can be modified by incubating the chromatin and the transcription extract for 30 min at room temperature after the addition of the transcription extract in step 1. The rNTP mix is then added, and the standard procedure is followed thereafter.

Primer extension is used to quantify transcription because it provides a sensitive measure of correctly initiated transcripts. For experiments involving RNA polymerase II, it is also important to determine that the transcripts are sensitive to α-amanitin, which specifically inhibits RNA polymerase II at low concentrations (0.5–5 μg/ml). With both *Drosophila* (12) and HeLa (13) nuclear extracts, the amount of transcription from activated templates is similar to the amount of transcription from naked DNA. An example of transcriptional analysis of chromatin is shown in *Figure 2*.

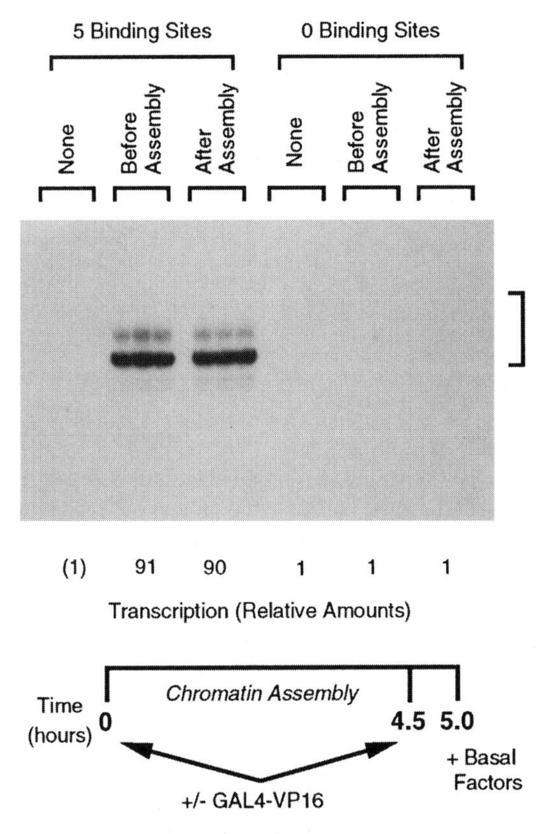

Figure 2. GAL4-VP16 activates transcription from histone H1-containing chromatin templates when added either before or after completion of chromatin assembly in the presence of the S-190 extract. H1-containing chromatin was assembled with either pGIE-0 or pIE-0, which contain five or zero GAL4 binding sites, respectively, upstream of the minimal adenovirus E4 promoter. Where indicated, GAL4-VP16 was added to the assembly reactions either before (0 hours) or after (4.5 hours) chromatin assembly. The resulting samples (containing 75 ng DNA) were subjected to *in vitro* transcription analysis as described in *Protocol 4*. In this experiment, GAL4-VP16 activates transcription 90-fold when added before or after assembly, but only in the presence of specific binding sites.

Protocol 4. Transcription of chromatin and primer extension analysis

Equipment and reagents

- 15 μl chromatin (containing 75 ng DNA), freshly prepared (*Protocol 1*)
- Transcription extract (soluble nuclear fraction, SNF)[a]
- rNTP mix. For 10 μl/sample: 5 mM of each of the four rNTPs (5 μl); 1 M Hepes, K^+, pH 7.6 (0.8 μl); 0.5 × 0.1 M HEMG (4.2 μl). Prepare immediately before use.
- 0.5 × 0.1 M HEMG: 12.5 mM Hepes, K^+, pH 7.6, containing 0.05 mM EDTA, 6.25 mM $MgCl_2$, 5% (v/v) glycerol, 50 mM KCl
- PvOH–PEG mix: 16.8 mM Hepes, K^+, pH 7.6, containing 0.11 M KCl, 3.3% (w/v) polyvinyl alcohol (MW 10 000; Sigma, P-8136), 3.3% (w/v) polyethylene glycol compound (MW 15 000–20 000; Sigma, P-2263)
- Transcription stop solution–proteinase K (*Protocol 2*)
- Phenol:chloroform:isoamyl alcohol (25:24:1 (v/v/v))
- Labelled oligonucleotide primer (Typically, a 28 nt oligonucleotide is 5′ end-labelled with T4 polynucleotide kinase and (γ-^{32}P)ATP (e.g. as in ref. 24). The optimal annealing conditions for step 8 will vary depending upon the length and G+C content of the oligonucleotide.)
- 0.3 M sodium acetate
- Annealing mix: 10 mM Tris–HCl, pH 7.8, containing 1 mM EDTA, 0.25 M KCl, 0.83 nM labelled oligonucleotide primer
- Extension mix. Add the following, in order: 1 M Tris–HCl, pH 8.3 (2.5 ml); 0.1 M $MnCl_2$ (500 μl); 2 ml 2.5 mg/ml actinomycin D in water, 132 μl 0.1 M solution of each of the four dNTPs, 1 ml 0.5 M DTT. Add water to 40 ml final volume. Store in 1 ml aliquots at – 20 °C in the dark.
- RT–PE mix. Immediately before use, combine, on ice, 10 units reverse transcriptase (typically, about 0.5–1 μl) from Moloney Murine Leukemia Virus (Stratagene, 600084) with 40 μl extension mix (per reaction).
- Formamide loading buffer (FLB): 80% (v/v) formamide containing 10 mM EDTA, 1 mg/ml xylene cyanol, 1 mg/ml Bromophenol blue
- FLB–NaOH loading buffer: 2 vol. FLB + 1 vol. 0.1 M NaOH. Prepare immediately before use.
- 8% (w/v) polyacrylamide–urea sequencing gel
- Rotary concentrator (e.g. Speed-Vac)

Method

1. Add to a 1.5 ml microcentrifuge tube, in the order listed, the following reagents. Mix gently by flicking the tube after each addition.
   - PvOH–PEG mix 15 μl
   - chromatin 15 μl (containing 75 ng DNA)
   - transcription extract 10 μl
   - rNTP mix 10 μl
2. Incubate for 30 min at room temperature.
3. Add 100 μl transcription stop solution–proteinase K. Vortex. Incubate for 15 min at 37 °C. (Can store samples at − 20 °C after this step.)
4. Add 250 μl 0.3 M sodium acetate, and then extract with 400 μl phenol:chloroform:isoamyl alcohol. Mix by vortexing. Centrifuge for 5 min at room temperature 15 000 *g*, in a microcentrifuge.
5. Transfer the aqueous phase to a 1.5 ml microcentrifuge tube containing 1 ml ethanol. Mix by vortexing. Centrifuge for 15 min at 15 000 *g* at room temperature in a microcentrifuge.
6. Remove the supernatant, and dry the pellet in a rotary concentrator (e.g. Speed-Vac). (Can store samples at − 20 °C after this step.)

7. Resuspend the pellet in 10 μl annealing mix. Incubate for 2 min at 70–75°C.
8. Incubate for 40 min at 58°C. (The optimal annealing conditions may vary depending upon the oligonucleotide used.)
9. Allow the samples to cool to room temperature. Add 40 μl RT–PE mix, and mix by flicking the tube. Incubate for 40 min at 37°C.
10. Add 225 μl of ethanol, mix by vortexing, and centrifuge for 15 min at 15 000 *g* in a microcentrifuge at room temperature.
11. Remove the supernatant, and dry the pellet in a rotary concentrator.
12. Resuspend the pellet in 5 μl FLB–NaOH loading buffer, and subject the DNA to electrophoresis on a standard 8% (w/v) polyacrylamide–urea sequencing gel.

[a] See ref. 23.

5. Analysis of chromatin template structure: mechanisms of transcriptional activation

Many mechanisms for the activation of transcription in the context of the native chromatin template have been proposed. For example, sequence-specific DNA transcription factors may bind to chromatin and reconfigure the nucleosomes in a manner that allows the binding of the basal transcriptional machinery (i.e. TFIID, TFIIB, etc.). In this model, the sequence-specific factors might directly rearrange or unfold nucleosomes, or possibly recruit other factors that perform such tasks. In addition, covalent modification of the histones (such as acetylation or phosphorylation) and the binding of histone H1 may have important roles in the control of transcriptional activity. Direct or indirect interactions between the sequence-specific transcription factors and the basal factors might also be required to counteract chromatin-mediated repression and to facilitate initiation and/or elongation of transcription. To address some of these issues, we have included the following experiments with which the mechanisms of transcription from chromatin templates could be examined.

5.1 Is a transcription factor bound to the chromatin?

DNase I-primer extension footprinting can be used to determine occupancy of factor binding sites in chromatin. Proteins bind to chromatin with varying affinities, and sequence-specific factors that are bound to chromatin may activate transcription with different potencies. For example, if a factor does not activate transcription, it may not be able to bind to the chromatin; or alternatively, the factor might be bound to the template, but not be capable of mediating other processes, such as the reconfiguration of chromatin structure or the assembly of the pre-initiation complex. *Protocol 5* describes a method to determine the binding of sequence-specific DNA binding proteins to chromatin. This method can be easily performed on a small scale, and can thus be used with a sample of chromatin that is also to be subjected to transcriptional and structural analyses. *Figure 3* shows an example of primer extension footprinting with chromatin templates.

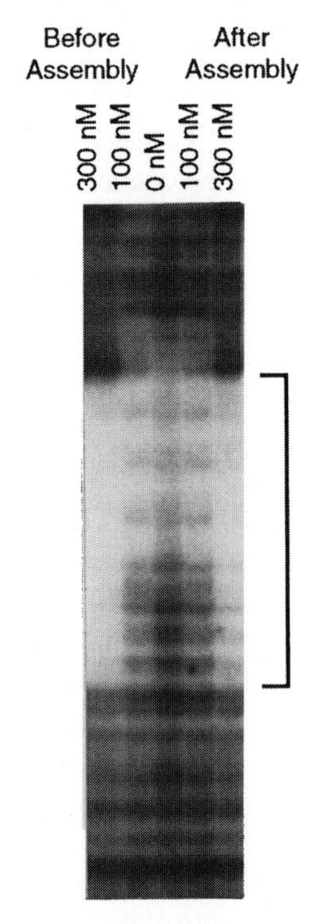

Figure 3. DNase I footprinting analysis. H1-containing chromatin was reconstituted in the presence of the indicated concentrations of GAL4-VP16 added either before or after chromatin assembly, and the resulting samples were subjected to DNase I digestion and primer extension analysis as described in *Protocol 5*. In this experiment, GAL4-VP16 binds chromatin before and after assembly with similar affinity.

Protocol 5. DNase I-primer extension footprinting of chromatin[a]

Equipment and reagents

- 50 μl chromatin (containing 250 ng DNA), freshly prepared (*Protocol 1*)
- 2.5 mg/ml DNase I (Worthington DPFF grade) stock solution in water. Store at −70°C in 10 μl aliquots. (Use each aliquot once and then discard.)
- DNase I dilution. Dilute DNase I stock 1:10 (v/v) in buffer R (*Protocol* 1) immediately before use.
- 10 × Vent buffer: 0.2 M Tris–HCl, pH 8.7, containing 0.1 M KCl, 0.1 M NH_4SO_4, 1% (v/v) Triton X-100
- RNase (DNase-free) (*Protocol 2*)

- MNase stop solution (*Protocol 3*)
- Transcription stop solution–proteinase K (*Protocol 2*)
- Phenol:chloroform:isoamyl alcohol (25:24:1 (v/v/v))
- 2.5 M ammonium acetate and 3 M sodium acetate
- Labelled oligonucleotide primer (Typically, a 28 nt oligonucleotide is 5′ end-labelled with T4 polynucleotide kinase and [γ-^{32}P]ATP (e.g. as in ref. 24). The optimal annealing conditions for step 11 will vary depending upon the length and G+C content of the oligonucleotide.)
- FLB–NaOH loading buffer (*Protocol 4*)
- Vent extension mix: for each sample, combine in order, on ice, immediately before use: water (32.5 μl); 10 × Vent buffer (5 μl), 1 μl 10 mM of each of the four dNTPs, 0.5 μl 0.1 M $MgSO_4$, 0.5 μl (= 16.6 fmol) labelled oligonucleotide primer, and 0.5 μl (= 1 unit) $Vent_R$ (exo$^-$) DNA polymerase (New England Biolabs, 257). (In our hands, $Vent_R$ (exo$^-$) works much more efficiently than the standard $Vent_R$ polymerase for primer extension analysis.)
- 8% (w/v) polyacrylamide–urea sequencing gel
- Rotary concentrator (e.g. Speed-Vac)
- Thermocycling apparatus

Method

1. Add 50 μl chromatin (containing 125 ng DNA) to a microcentrifuge tube containing 1.5 μl 0.1 M $CaCl_2$. Mix immediately by flicking the tube gently.
2. Add 5 μl of the DNase I dilution. Mix immediately by flicking the tube gently. Start multiple samples consecutively, 15 sec apart. Incubate for 1 min at room temperature.
3. Add 5 μl of MNase stop solution. Mix immediately by flicking the tube gently. Stop multiple samples consecutively, 15 sec apart. Incubate for approximately 15 min at room temperature. (Samples can be stored at – 20 °C after this step.)
4. Add 100 μl of transcription stop solution–proteinase K. Incubate for 15 min at 37 °C.
5. Extract with 150 μl of phenol:chloroform:isoamyl alcohol.
6. Transfer the aqueous phase to a 0.65 ml microcentrifuge tube containing 15 μl 2.5 M ammonium acetate and 340 μl ethanol.
7. Vortex, and centrifuge for 15 min at 15 000 *g* at room temperature in a microcentrifuge.
8. Carefully remove the supernatant, and dry the pellet in a rotary concentrator
9. Resuspend the pellet in 10 μl of water.
10. Add 40 μl of Vent extension mix at 4 °C. Vortex gently.
11. Set up the tubes in a thermocycling apparatus. Denature for 4 min at 95 °C, and then perform eight cycles of the following: 1 min denaturing at 94 °C; 3 min annealing at 55 °C;1 min extension at 74 °C.
12. Put the samples on ice. Add 5 μl 3 M sodium acetate and 120 μl ethanol, and precipitate as before (steps 7 and 8).
13. Resuspend in 9 μl FLB–NaOH loading buffer. Load 4 μl onto a standard 8% (w/v) polyacrylamide–urea sequencing gel.

[a] See *Protocol 2*, Chapter 4.

5.2 Does a transcription factor change the chromatin structure?

Disruption of the periodicity of an array of nucleosomes in a specific region is analysed by probing Southern transfers of micrococcal nuclease ladders

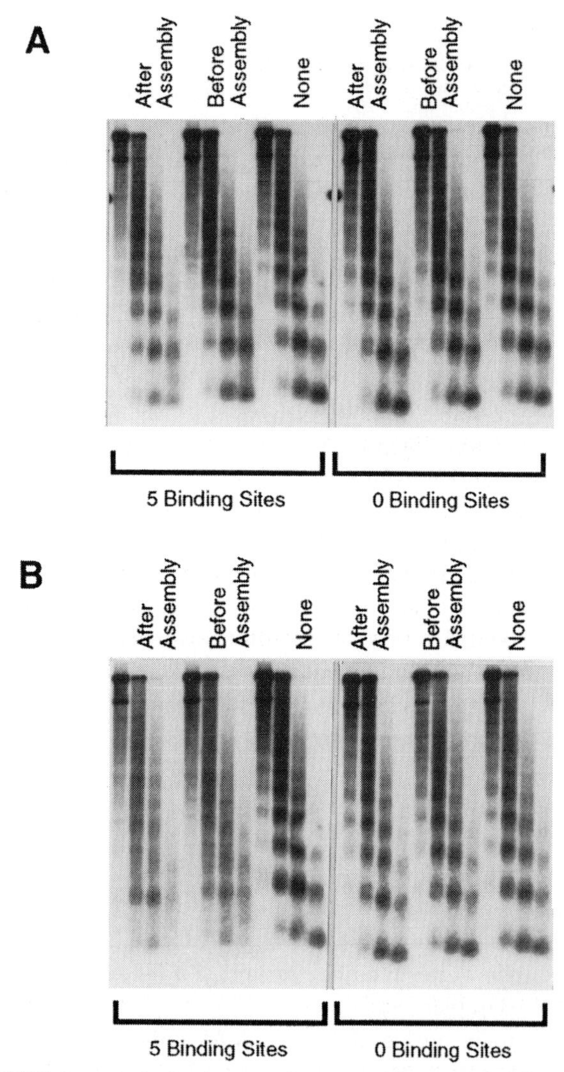

Figure 4. GAL4-VP16 disrupts the nucleosomal array at the promoter. Micrococcal nuclease-digested chromatin was analysed as described in *Protocol 6* and sequentially hybridized to oligonucleotide probes that correspond to sequences either approximately 900 bp upstream of the RNA start site (A) or between the GAL4 sites and the RNA start site (B). In this experiment, GAL4-VP16 disrupts the periodic nucleosomal array only at the promoter, only when GAL4 binding sites are present.

(*Protocol 6*) (12, 13, 25). These experiments are performed in the absence of rNTPs and transcription extract to exclude the possibility that any observed changes in chromatin structure were a consequence of transcription of the template. In a typical experiment, chromatin is assembled in the presence or absence of a transcriptional activator, digested with micrococcal nuclease, analysed by agarose gel electrophoresis (recommended: gel is stained with ethidium bromide and photographed), transferred to nitrocellulose, and hybridized to probes specific for the region of interest. If the factor does not cause any disruption of the regularity of the nucleosomal ladder in the region of the probe, then the resulting autoradiogram reveals a standard micrococcal digestion ladder (*Figure 4A*). On the other hand, if there is disruption of the nucleosome structure in the region of the probe, then a micrococcal nuclease ladder is not observed (*Figure 4B*).

In the study of a promoter, it is useful to use a probe that is in the vicinity of the transcription factor binding sites (where disruption of the micrococcal ladder might be observed) in conjunction with a probe corresponding to a distal location that is about 1 kb away from the factor binding sites (with which disruption of the micrococcal ladder might not be seen). The results obtained with the distal probe serve as a control, because the observation of a regular micrococcal ladder (as in *Figure 4A*) demonstrates that the disruption of the chromatin structure in the vicinity of the factor binding sites (as in *Figure 4B*, for the samples with binding sites and transcriptional activator) was due to a local effect, rather than a general, non-specific effect that occurs throughout the plasmid. As an independent test of a general inhibition of chromatin assembly, the micrococcal ladders that are seen in these Southern blot experiments can also be compared with the ladders that were observed upon ethidium bromide staining of the bulk DNA in the agarose gels prior to the transfer of the DNA fragments to nitrocellulose.

Protocol 6. Nucleosomal array disruption assay

Equipment and reagents

- 0.2 μm nitrocellulose filter (Micron Separations, WP2HY320F5)
- Denaturing solution: 0.5 M NaOH, 1.5 M NaCl
- Neutralizing buffer: 0.5 M Tris–HCl, pH 7.5, 1.5 M NaCl
- 20 × SSPE: 3 M NaCl, 0.2 M NaH_2PO_4, 20 mM EDTA. Adjust to pH 7.4 with NaOH, and then autoclave.
- Non-fat dried milk (e.g. Carnation)
- 0.1% (w/v) SDS
- Standard Southern transfer stack[a]
- 3MM paper (Whatman)
- Labelled oligonucleotide probe. (Typically, a 28 nt oligonucleotide is 5′ end-labelled with T4 polynucleotide kinase and [γ-^{32}P]ATP[a]. The hybridization and wash conditions, as described, will vary depending upon the length and G+C content of the oligonucleotide.)
- Vacuum oven
- Heat-seal hybridization bag
- Plastic wrap
- Autoradiography equipment

A. *Transfer*

1. Perform micrococcal nuclease analysis of chromatin exactly as

Protocol 6. *Continued*

described in *Protocol 3* (except that the gel is not destained in step 11 of that protocol).

2. Denature the DNA by incubating the gel twice, for 15 min each time, in a solution containing 0.5 M NaOH and 1.5 M NaCl. At the same time, soak a nitrocellulose filter in water.
3. Rinse the gel with water, and neutralize the gel by incubating twice, 20 min each time, in the neutralizing buffer. At the same time, transfer the nitrocellulose filter to 20 × SSPE.
4. Capillary transfer by using 20 × SSPE and a standard Southern transfer stack[a].
5. Place the filter between two dry sheets of 3MM paper (Whatman), and bake for 1 h at 80 °C in a vacuum oven. Use the dried filter immediately or store at room temperature.

B. *Hybridization*

1. Place the filter in a heat-seal hybridization bag. Add 20 ml of 6 × SSPE containing 0.25% (w/v) dried milk (for a filter with dimensions between 10 × 10 cm and 20 × 20 cm).
2. Pre-hybridize for 1 h at 67.5 °C.
3. Add 1.5 pmol of phosphorylated oligonucleotide probe (stored in TE buffer at 0.1 pmol/μl) to the contents of the bag, and reseal the bag.
4. Incubate for 3–24 hours at 67.5 °C.
5. Cut open one corner of the bag, and pour out the hybridization mix.
6. Rinse the filter once in the hybridization bag with 10 ml of 6 × SSPE (buffer at room temperature).
7. Add 20 ml of 6 × SSPE to the hybridization bag, reseal the bag, and incubate for 10 min at 67.5 °C to wash the filters.
8. Rinse the filter once again in the hybridization bag with 10 ml of 6 × SSPE (buffer at room temperature).
9. Remove the filter from the hybridization bag, drain the excess liquid from the filter, wrap the filter with plastic wrap, and autoradiograph.
10. To strip the labelled probe from the filter, perform the following. Heat a solution of 0.1% (w/v) SDS to a boil, then remove it from the heat. Place the filter in the heated SDS solution, shake the mixture for approximately 30 min, and then rinse the filter with water. Air-dry the filter if desired.
11. Store the filters damp in plastic wrap at room temperature, or strip and store dry.

[a] See ref. 24.

5.3 Are there nucleosomes positioned on the template?

The positioning of nucleosomes can be examined by indirect end-labelling after digestion of the chromatin with micrococcal nuclease (*Protocol 7*) (Chapter 8 and refs 26, 27) or MPE–Fe(II) (20) (see also the discussion and diagram in ref. 1). An aliquot of chromatin is lightly digested with micrococcal nuclease, and the resulting DNA is extracted, digested with a restriction enzyme (whose recognition site is located 0.5–2 kb from the region of interest), subjected to electrophoresis on an agarose gel, transferred to nitrocellulose, and hybridized to an oligonucleotide probe that binds to the DNA segment of interest at a location adjacent to the restriction enzyme cleavage

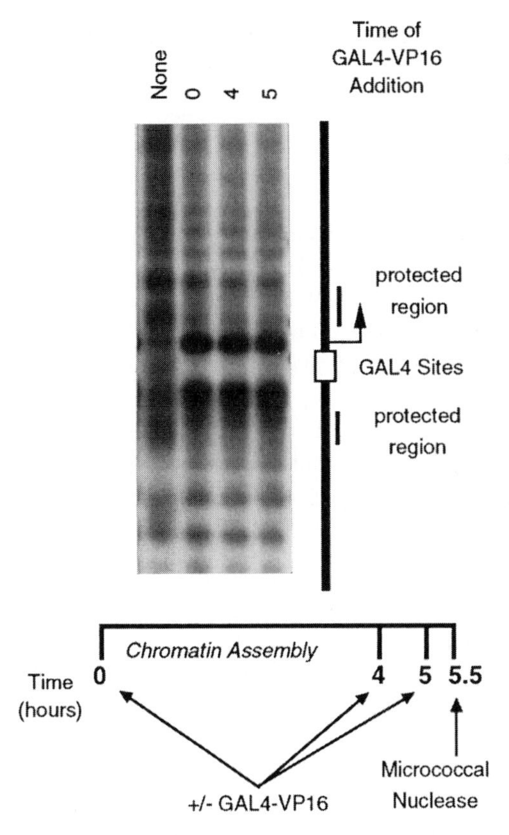

Figure 5. GAL4-VP16 reconfigures the chromatin structure when added either before or after chromatin assembly. Chromatin was assembled with pGIE-0 DNA (which contains five GAL4 binding sites immediately upstream of the TATA box of the adenovirus E4 minimal promoter) in the absence of histone H1, and, where noted, GAL4-VP16 was added to the reactions either before (0 hours) or after (4 or 5 hours) chromatin assembly. The resulting samples were then subjected to indirect end-labelling analysis as described in *Protocol 7*. In this experiment, GAL4-VP16 directs the same reconfiguration of the chromatin template when added before or after assembly.

site. In practice, the chromatin can be digested with two restriction enzymes on opposite sides of the region of interest, and probed sequentially from both sides. An example of an indirect end-labelling experiment is shown in *Figure 5*.

Protocol 7. Indirect end-labelling

Equipment and reagents

- 100 μl chromatin (containing 500 ng DNA), freshly prepared (*Protocol 1*)
- MNase dilutions. For each sample of chromatin, dilute MNase stock (*Protocol 3*) in buffer R (*Protocol 1*) to give 1.3 U/ml and 4 U/ml MNase. Make immediately before use.
- Appropriate restriction enzymes
- 2.5 M ammonium acetate
- Agarose gel (*Protocol 3*)
- Labelled oligonucleotide probe. (Typically, a 28 nt oligonucleotide is 5′ end-labelled with T4 polynucleotide kinase and [γ-^{32}P]ATP[a]. The optimal hybridizing and wash conditions will vary depending upon the length and G+C content of the oligonucleotide.)
- Rotary concentrator (e.g. Speed-Vac)
- Equipment and reagents for Southern blotting and hybridization (*Protocol 6*)

Method

1. Digest chromatin (50 μl of chromatin containing 250 ng of DNA), as described for the micrococcal nuclease analysis of chromatin (*Protocol 3*, steps 1–7), but use the MNase dilutions shown above. Thus, for each chromatin sample digest 50 μl with the 1.3 U/ml MNase solution and separately digest another 50 μl of chromatin with the 4 U/ml MNase solution.
2. Resuspend the nucleic acids in water, and digest with the appropriate restriction enzyme(s) in a final volume of 20 μl for 4 hours at 37 °C.
3. Precipitate with 2 μl 2.5 M ammonium acetate and 55 μl ethanol. Centrifuge for 15 min at 15000 *g* at room temperature in a microcentrifuge.
4. Carefully remove the supernatant, and dry the pellet in a rotary concentrator.
5. Resuspend the pellet, and resolve the DNA fragments on an agarose gel, as described in *Protocol 3*, steps 8–10.
6. Perform a Southern transfer, and hybridize the filter with an oligonucleotide probe adjacent to the restriction site used in step 2, as described in *Protocol 6*.

[a]See ref. 24.

6. Summary and prospects

The strategies and techniques described in this chapter represent the early stages in the study of the role of chromatin structure in the regulation of gene expression, and there are many conceptual and technical barriers that must be surmounted for further progress to be achieved. Some of these problems include methodology for high-resolution determination of nucleosome and chromatin structure; identification and characterization of the factors and mechanisms involved in chromatin assembly and reconfiguration; and elucidation of higher-order chromatin structures, such as heterochromatin. The pursuit of these and other such goals will inevitably lead to many important insights into the fundamental mechanisms by which genes are regulated in their natural context.

Acknowledgements

We thank Catherine George, Thomas Burke, and Mike Bulger for helpful suggestions on the manuscript. J. T. Kadonaga is supported by grants from the National Institutes of Health, National Science Foundation, and the Council for Tobacco Research.

References

1. Paranjape, S. M., Kamakaka, R. T., and Kadonaga, J. T. (1994). *Annu. Rev. Biochem.*, **63**, 265.
2. Wallrath, L. L., Lu, Q., Granok, H., and Elgin, S. C. R. (1994). *BioEssays*, **16**, 165.
3. Wolffe, A. P. (1994). *Curr. Opin. Genet. Dev.*, **4**, 245.
4. Workman, J. L. and Buchman, A. R. (1993). *Trends Biochem. Sci.*, **18**, 90.
5. Rhodes, D. and Laskey, R. A. (1989). In *Methods in enzymology* (ed. P. M. Wassarman and R. D. Kornberg), Vol. 170, pp. 575–84. Academic Press, New York.
6. Stein, A. (1989). In *Methods in enzymology* (ed. P. M. Wassarman and R. D. Kornberg), Vol. 170, pp. 585–602. Academic Press, New York.
7. Shimamura, A., Jessee, B., and Worcel, A. (1989). In *Methods in enzymology* (ed. P. M. Wassarman and R. D. Kornberg), Vol. 170, pp. 603–11. Academic Press, New York.
8. Sealy, L., Burgess, R. R., Cotten, M., and Chalkley, R. (1989). In *Methods in enzymology* (ed. P. M. Wassarman and R. D. Kornberg), Vol. 170, pp. 612–29. Academic Press, New York.
9. Becker, P. B. and Wu, C. (1992). *Mol. Cell. Biol.*, **12**, 2241.
10. Kamakaka, R. T., Bulger, M., and Kadonaga, J. T. (1993). *Gen. Dev.*, **7**, 1779.
11. Bulger, M. and Kadonaga, J. T. (1994). In *Methods in molecular genetics* (ed. K. W. Adolph), Vol. 5, pp. 241–62. Academic Press, San Diego.
12. Pazin, M. J., Kamakaka, R. T., and Kadonaga, J. T. (1994). *Science*, **266**, 2007.
13. Pazin, M. J., Sheridan, P. L., Cannon, K., Cao, Z., Keck, J. G., Kadonaga, J. T., and Jones, K. A. (1996). *Gen. Dev.*, **10**, 37.
14. Bates, D. L. and Thomas, J. O. (1981). *Nucl. Acids Res.*, **9**, 5883.
15. Barton, M. B., Madani, M., and Emerson, B. (1993). *Gen. Dev.*, **7**, 1796.
16. Croston, G. E., Lira, L. M., and Kadonaga, J. T. (1991). *Protein Expr. Purif.*, **2**, 162.

17. Shimamura, A., Tremethick, D., and Worcel, A. (1988). *Mol. Cell. Biol.*, **8**, 4257.
18. Camerini-Otero, R. D. and Felsenfeld, G. (1977). *Nucl. Acids Res.*, **4**, 1159.
19. Noll, M. and Kornberg, R. D. (1977). *J. Mol. Biol.*, **109**, 393.
20. Cartwright, I. L. and Elgin, S. C. R. (1989) In *Methods in enzymology* (ed. P. M. Wassarman and R. D. Kornberg), Vol. 170, pp. 359–68. Academic Press, New York.
21. Wang, T. and Allis, C. D. (1993). *Mol. Cell. Biol.*, **13**, 163.
22. Kamakaka, R. T., Tyree, C. M., and Kadonaga, J. T. (1991). *Proc. Natl Acad. Sci. USA*, **88**, 1024.
23. Kamakaka, R. T. and Kadonaga, J. T. (1994). In *Methods in cell biology* (ed. L. S. B. Goldstein and E. A. Fyrberg), Vol. 44, pp. 225–35. Academic Press, San Diego.
24. Sambrook, J., Fritsch, E. F., and Maniatis, T. (1989). *Molecular cloning: a laboratory manual* (2nd edn). Cold Spring Harbor Press, New York.
25. Tsukiyama, T., Becker, P. B., and Wu, C. (1994). *Nature*, **367**, 525.
26. Wu, C. (1980). *Nature*, **286**, 854.
27. Nedospasov, S. A. and Georgiev, G. P. (1980). *Biochem. Biophys. Res. Commun.*, **92**, 532.

10

Assembly of chromatin and nuclear structures in *Xenopus* egg extracts

GENEVIÈVE ALMOUZNI

1. Introduction

The structural organization of DNA within the eukaryotic nucleus is tightly defined to accommodate the functional properties of a given cell type. Nuclear functions such as replication and transcription operate in a highly regulated fashion using chromatin as a template. During every cell cycle, each nucleus has to replicate itself, requiring duplication not only of DNA but also of the individual chromosomes together with all their structural and regulatory features. Chromatin organization is known to influence gene expression (1) and its remodelling during replication and cell cycling can affect the transcription process. To study the relationship between the structural organization of DNA and macromolecular events such as replication and transcription, an adequate system is required, accurately reproducing both structural and functional features of the eukaryotic cell.

The recent development of various types of extracts derived from *Xenopus* eggs has been a major boost in our approach to the assembly of a nucleus, its structure and function. The pioneering work of Lohka and Masui (2) to produce egg extracts in which nuclear reconstitution and spindle assembly could take place, together with other earlier *in vitro* approaches to the study of chromatin assembly (Chapters 8 and 9 and ref. 3), has led to the most recently refined use of systems to follow both structural and functional properties of the DNA within the nucleus. This was possible because of the major advantages related to this biological system. *Xenopus* eggs can be obtained in large quantities. Due to the accumulation of material during oogenesis, they provide an invaluable biochemical source of material, including both structural and regulatory factors (4–9). This chapter presents the main characteristics for the preparation of egg extracts and their fractionation. I will report on the assembly of nuclear structures, and also explain how to use these extracts to study transcription or replication.

2. Obtaining and preparing eggs

The quality of an egg extract is strictly dependent on the quality of the eggs. It is essential to maintain *X. laevis* female frogs in adequate conditions as previously described (10). By hormonally inducing them to lay eggs, more than 1000 eggs are routinely obtained from an individual frog. Note that the number of eggs and their quality is usually poor in summer. The quality of the eggs is examined under a dissecting microscope. Normal-looking eggs present a dark animal pole with a white spot on the top, indicative of maturation, separated from the clear vegetal pole by a thin white zone. They are surrounded by a jelly coat. Eggs from different frogs are usually collected and treated separately, especially during the early treatment involving the removal of the jelly coat. Indeed, inter-animal variations in the amount of jelly present are often observed, and thus specific conditions must be adjusted for each case. The jelly is held together by disulfide bonds which can be broken by reducing agents such as cysteine. Usually, the eggs collected for an extract preparation have accumulated during the night. However, freshly laid eggs can be useful if fertilization is considered. In this case, the frogs are squeezed gently and the eggs can be collected and processed immediately after laying.

Protocol 1. Obtaining eggs from *Xenopus* females by hormone stimulation

Equipment and reagents

- 3 female frogs (CRBM-CNRS). After use keep the injected frogs at least 3 months before injecting them again.
- Pregnant mare serum gonadotropin (PMSG, Intervet)
- Human chorionic gonadotropin (HCG, Sigma)
- 3 buckets (5-litre capacity) with a lid and a Plexiglas tray with holes (1 cm diameter) to fit into the bucket. Put each frog to lay in a different bucket so as not to mix the eggs.
- 1 ml syringes, 27-gauge needles
- 0.1 M NaCl in deionized water
- L-cysteine hydrochloride monohydrate (Sigma, C-7880): 200 ml of a 2% cysteine solution in 0.1 M NaCl (adjust to pH 7.8 using 10 M NaOH). Prepare just before use.
- Glassware: beaker (250 ml or 500 ml), large Petri dish. (Avoid plastic because the eggs tend to stick to it.)
- Glass Pasteur pipette with a cut and fire-polished tip (3 mm diameter)

A. *PMSG injection*

This procedure is to accelerate the growth of oocytes; omit if large females are used.

1. Inject 0.5 ml of a PMSG solution (200 U/ml in sterile water) subcutaneously into the lymphatic sacs in the back. For each frog use a different needle as it can be clogged by secretions from frog skin.
2. Keep the frogs for 3–10 days maximum. A longer time might induce apoptotic extracts as recently reported[a].

B. *HCG injection*

1. Dilute the HCG solution to 1500 U/ml in sterile water. Store the solution at 4°C for a maximum of 1 week, or freeze at −20°C and store for up to 3 months.
2. Inject 750 U for a medium-sized frog, 1000 U for a large frog. Excess hormone may give batches of eggs of poor quality.
3. Keep the frogs individually in each of the buckets filled with 3 litres of 0.1 M NaCl (the eggs activate spontaneously in pure water). The tray at the bottom of the bucket prevents frogs from spoiling their eggs. If maintained at 22°C, the frogs will start laying after about 6 h. Avoid higher temperatures or the eggs will deteriorate; use lower temperatures to slow egg laying.

C. *Jelly removal*

1. Transfer the eggs from the bucket to a 500 ml beaker; wash them extensively with clean 0.1 M NaCl solution and finally rinse them with distilled water. Allow the jelly coat to swell in distilled water for about 5 min. When the eggs are stuck, discard all the water. Expect about 30 ml of eggs.
2. Cover the eggs with the cysteine solution and swirl gently until the jelly is dissolved, the eggs are closely packed (about 10 ml) and the medium becomes turbid (around 3–4 min). Do not leave longer than 5 min.
3. Wash the eggs three times with 100 ml of 0.1 M NaCl. Transfer the eggs to a glass Petri dish and sort under a dissecting microscope. Because the eggs are fragile at this stage, proceed immediately to the preparation of extracts.

[a] See ref. 11.

3. The egg extracts

3.1 The different types of extracts

Different types of extracts can be obtained according to:

(a) The speed of the centrifugation used to prepare the extracts—mainly two types can be distinguished: a low-speed extract (LSE) and a high-speed extract (HSE) (see *Figure 1*). The content of these extracts is different and determines their ability to reproduce cellular events.
(b) The state of the egg with respect to the cell cycle—the extracts obtained have been distinguished as: M-phase extract or interphase extract and cycling extracts. Their properties mimic those observed *in vivo* for the corresponding time in the cell cycle.

Xenopus eggs are physiologically arrested in metaphase of the second meiotic division. At this stage they contain high levels of maturation (or mitosis) promoting factor (MPF) as well as cytostatic factor (CSF). In the initial step of the extract preparation the cells are simply broken open by centrifugation at low speed (10 000 *g*). This lysis results in a transient release of Ca^{2+} from intracellular stores, as seen at fertilization following sperm penetration. This leads to the release of the metaphase block. Therefore, to prepare an M-phase extract, also called CSF-arrested extract, lysis is achieved in the presence of EGTA which chelates Ca^{2+} and thus prevents exit from M phase (12). These extracts are commonly called mitotic extracts, although they correspond, in fact, to metaphase of the second meiotic division. In contrast, in the absence of EGTA in the lysis buffer, the extract will not be of the M-phase type. Ca^{2+} release can promote the proteolysis of cyclin, which causes the destruction of MPF, allowing exit from M phase and entry into S phase. However, if cyclin synthesis is allowed, cell-cycle progression may continue. When high-speed centrifugation (150 000 *g*) is performed on the low-speed extract, the supernatant is devoid of polyribosomes and protein synthesis does not occur. This will keep the extract in S phase. High-speed extracts made according to this method have been used extensively to study chromatin assembly (6). This high-speed centrifugation also separates the membranous fraction from the cytosolic fraction. Hence, the extracts are unable to assemble nuclei and replicate double-stranded DNA.

To obtain a large amount of eggs at different times in the cell cycle, the most physiological approach would be synchronous fertilization. This is feasible, however, only on a small scale. Instead, one can 'activate' the eggs to induce a Ca^{2+} release (by an electric shock or by the use of an ionophore). Under these conditions the low-speed extract obtained is efficient in protein synthesis. This protein synthesis generates cyclin molecules that ensure the oscillation between S phase and M phase. These are cycling extracts. Developed by both Hutchison *et al.*(13) and Murray and Kirschner (14), these extracts can undergo multiple cell cycles *in vitro*. Cyclin resynthesis can be prevented by the addition of cycloheximide to prevent protein synthesis; in this case the extract is called an interphase extract. Alternatively, starting from activated eggs, it is possible to prepare an extract representing different phases of the cell cycle without blocking protein synthesis with cycloheximide (9). In this case, the extracts are centrifuged at the higher speed (150 000 *g*). Low-speed extracts contain membrane vesicles that can be used to reconstitute nuclei and initiate replication on double-stranded DNA (15). These extracts can be further fractionated by high-speed centrifugation. The membranous fraction can be separated from the cytosolic fraction. Both of these fractions can be stored separately, and recombination will restore the properties of the low-speed extract, with respect to nuclear assembly and DNA replication, but not for protein synthesis.

3.2. Preparation of egg extracts

As the preparation of M-phase extracts has been described in detail elsewhere (9), our attention will focus on cycling extracts or interphase extracts. It is interesting to note that M-phase extracts can be converted to interphasic extracts by simply adding calcium (16).

3.2.1 The cycling extract

The property of the extract to go through multiple cycles relies on its ability to perform protein metabolism, synthesis, and degradation. A low-speed centrifugation of the eggs will break them and stratify their contents. The intermediate brownish fraction called LSE, enriched in ribosomal component, is efficient for protein synthesis. For the best efficiency, the eggs are usually activated. It has been reported that in the absence of activation, the extracts go from interphase to mitosis and then stop (8). Electric shock appeared more reproducible in our hands than ionophore treatment. We also found that a better activation is obtained when starting from squeezed eggs compared to eggs laid during the night. The buffer used to prepare these extracts differs between the various laboratories (13, 14), and this may be of importance for specific purposes. Here, I will describe the one technique we are currently using to follow DNA replication or nuclear formation. For additional information, see ref. 8.

Protocol 2. Preparation of a cycling extract

Equipment and reagents

- 1 × Barth: 88 mM NaCl, 2 mM KCl, 1 mM $MgCl_2 \cdot 6H_2O$, 15 mM Tris–HCl, pH 7.6, 0.5 mM $CaCl_2 \cdot 6H_2O$
- 1 × Barth, 4% Ficoll
- Activation chamber: the one used here is homemade. The chamber is 11.5 cm × 9.5 cm (inside), 6 cm high, made in Plexiglas. The bottom of the chamber is covered with a grid of stainless steel and this constitutes one of the electrodes. The lid of the chamber is attached to a grid of stainless steel (same size as the one at the bottom of the chamber) which penetrates into the chamber bringing the two electrodes 5 cm apart.
- Extraction buffer: 10 mM K-Hepes, pH 7.8, 70 mM K-acetate, 5% sucrose, 0.5 mM DTT. Make a 5 × stock solution without DTT and protease inhibitors, filter, then store at 4°C for about a week or in 50 ml aliquots at −20°C. Keep on ice before use.
- 1% agarose in 0.1 × Barth (100 ml)
- Power supply: to deliver 15 V AC to the chamber.
- 1 M DTT, protease inhibitors stock solutions: dissolve both leupeptin and pepstatin A to a concentration of 10 mg/ml in dimethylsulfoxide (DMSO) and store at −20°C.
- 1 ml Gilson pipette
- Centrifugation tubes and centrifuges and rotors: use a Sorvall HB4 rotor (or Beckman J-21) and 15 ml Corex tubes for the low-speed centrifugation. Pre-cool rotors and tubes. Set temperature of the centrifuge at 2°C.
- Clinical centrifuge
- Mineral oil (General Electric, Versilube F-50)
- Glass Pasteur pipette
- Liquid nitrogen

A. *Activation of the eggs by an electric shock*

1. Cover the bottom of the chamber with 0.5 cm of 1% agarose in 0.1 × Barth and fill the chamber with 0.1 × Barth to proceed for activation.

Protocol 2. *Continued*

2. Disperse the dejellied eggs (*Protocol 1C*) in the chamber as a monolayer over the agarose.
3. Close the lid of the chamber and give a 2 sec 15 V pulse.
4. Remove the lid. The pigment at the animal pole contracts within 5 min when activation is successful. The white spot on the top of the egg is no longer detectable. The egg is free to rotate due to the elevation of the fertilization membrane and turns the dark pole up according to gravity.
5. Transfer the activated eggs to a Petri dish containing 1 × Barth, 4% Ficoll and discard damaged or abnormal eggs.
6. Keep the eggs for 15 min in 1 × Barth, 4% Ficoll at 20°C.

B. *Low-speed crushing*

1. Wash the eggs with 1 vol. of ice-cold extraction buffer supplemented with 0.5 mM DTT, and protease inhibitors at 10 μg/ml.
2. Transfer them (with a glass Pasteur pipette with a cut tip) to a 15-ml, Corex centrifuge tube. Do not fill the tube above 2/3 of its volume. Remove the excess buffer using a 1 ml Gilson pipette to obtain extracts as concentrated as possible. Keep everything on ice from this time on.
3. Pack the eggs gently into a clinical centrifuge and spin for 10 sec at 200 *g*. Add mineral oil after an initial 5 min centrifugation at 200 *g* (Versilube F-50, General Electric) to lower the dilution even further. The buffer which is less dense than the oil, floats over it.
4. Centrifuge for 15 min at 10 000 *g* in a swinging-bucket rotor (Sorvall HB4 or Beckman J-21) at 2 °C. This causes the cell content to stratify (*Figure 1*). From top to bottom you will find: a yellow layer of lipids, a brownish-grey cytosolic intermediate layer, and yolk layer which is at the bottom just beneath the dark pigments. (For small volumes, it is possible to use an Eppendorf centrifuge at maximum speed.)
5. Use a Pasteur pipette to puncture the yellow layer of lipids. Take a clean pipette to go through the hole made with the first pipette and remove the intermediate viscous and turbid supernatant, but avoid taking the pigments and yolk when coming close to the bottom of the tube.
6. Use the extract fresh or supplement the extract with 5% glycerol for freezing.
7. Make drops of 15 μl in liquid nitrogen. Store the frozen drops at – 80 °C.

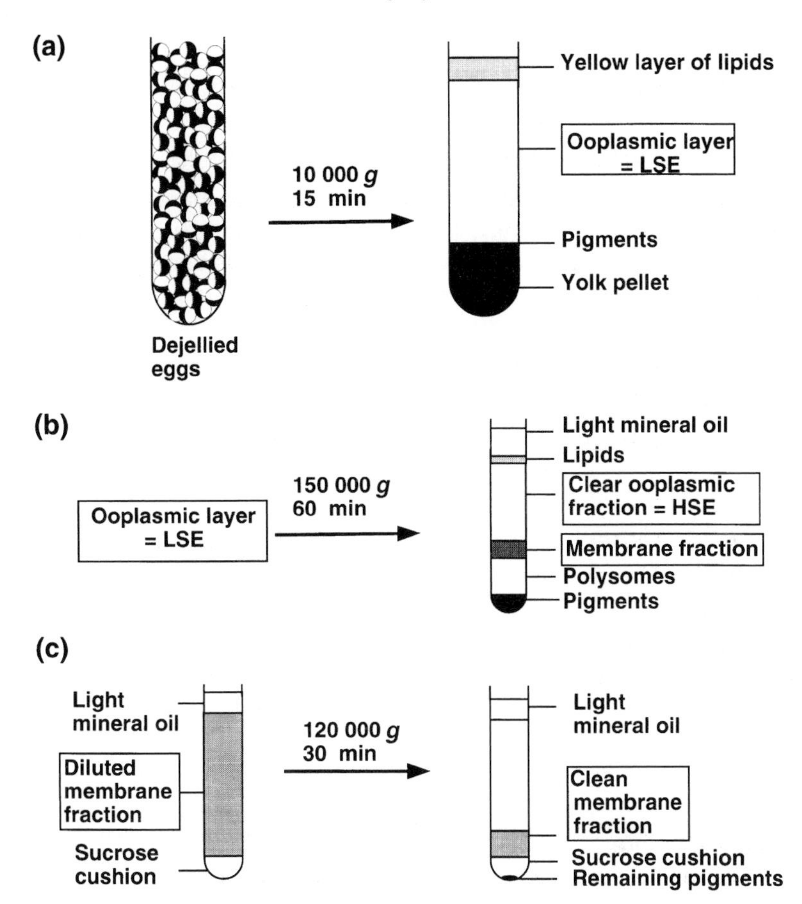

Figure 1. Preparation of egg extracts. (a) Preparation of a low-speed extract (LSE). Packed eggs are centrifuged at 10 000 *g* for 15 min according to *Protocol 2*. Distinct layers are obtained as shown. The ooplasmic layer is removed using a glass Pasteur pipette to pass through the yellow layer of lipids. (b) Preparation of a high-speed extract (HSE). The ooplasmic layer (LSE) obtained from (a) is transferred to an SW-50.1 tube and spun at 150 000 *g*. Stratification into different layers is as described. The clear ooplasmic fraction (HSE) is recovered through the wall of the tube with a 5 ml syringe, as well as the membrane fraction with another syringe. (c) The membrane fraction is then diluted as described in *Protocol 3*, and centrifuged again at 120 000 *g* for 30 min in an SW50.1 rotor. The clean membrane fraction is then recovered.

These extracts are to be used fresh, but they can be kept on ice for a few hours before use. Used under these conditions they will function as cycling extracts. Up to three or four cycles of the embryonic type can be followed with about 400 nuclei per μl extract. These extracts are very useful for following the dynamics of a structure as it progresses through the cell cycle. Any remaining extract can be frozen. However, because protein synthesis is sensitive to freezing, the frozen extract may, to some extent, lose its capacity to cycle. At the least, it will behave as an interphasic extract allowing nuclear assembly and DNA replication to be followed.

3.2.2 The interphasic extract

These extracts are useful for following events within an interphasic nuclei. The LSE, containing membrane precursors, can assemble chromatin within nuclei. The HSE, lacking the membrane precursors, cannot form nuclei but can assemble chromatin.

To prepare an interphasic extract it is possible to use the eggs just laid and rely on the 'autoactivation', due to the Ca^{2+} release when eggs are broken, to allow exit from M phase and entry into S phase (*Protocol 3*, Method 1). It is also possible to use activated eggs as a starting material—Method 2 in the same protocol.

Protocol 3. Preparation of interphase egg extract

Equipment and reagents

- Dejellied eggs (*Protocol 1*)
- Extraction buffer (*Protocol 2*)
- Centrifuge, rotor, and tubes for the low-speed centrifugation (*Protocol* 2)
- Ultracentrifuge (Beckman L70), a rotor (SW50.1), and 5 ml translucent tubes for the high-speed centrifugation
- 5 ml syringe and needles to pierce the centrifugation tubes. (We use the microlance 3 18G1/2 1.2 × 40 TWPM, Becton Dickinson.)
- Light mineral oil (Sigma)
- Cycloheximide (Sigma)
- 1.3 M sucrose in extraction buffer

Method 1: preparation of interphase egg extract using unactivated eggs

A. *Egg crushing and low-speed centrifugation*

1. Follow *Protocol 2*, but do not activate the eggs.
2. When using the LSE fresh, maintain the interphasic state by adding cycloheximide to the reaction mixture at 200 μg/ml. This is unnecessary if the extract is to be processed further for high-speed centrifugation.

B. *High-speed centrifugation and extract fractionation*

1. Transfer the LSE to the SW-50.1 ultraclear tube on ice. Fill the tube to the top with light mineral oil.
2. Centrifuge in a SW-50.1 rotor at 150 000 *g* for 1 h at 4 °C. The ultracentrifugation step separates the membranes and soluble components (*Figure 2*). Puncture the side of the tube to withdraw the membrane fraction and the clear soluble extract. Freeze the soluble extract as described in *Protocol 2*.
3. Resuspend the membrane fraction in cold extraction buffer and transfer to an ultraclear centrifuge tube containing 500 μl of 1.3 M sucrose in extraction buffer to form a cushion.

4. Pellet the membranes by centrifugation at 120 000 *g* for 30 min at 2°C in a Beckman ultracentrifuge.
5. Remove the supernatant and collect the membranes in the smallest volume of buffer. It should be about 1/10 the volume of the soluble fraction.
6. Supplement the membrane fraction with sucrose (0.5 M final comcentration) prior to freezing (5% glycerol can also be used).
7. Freeze 15 μl drops into liquid nitrogen and store the beads at − 80°C.

Method 2: preparation of interphase egg extract using activated eggs

1. Include cycloheximide (200 μg/ml) both before and after activation.
2. After 10–20 min incubation following activation of the eggs, proceed as described in *Protocol 2*.

The two types of extracts can then be used to study nuclear events as described in the next sections.

3.3 Labelling the extract

Labelling the extract can be very useful in chromatin and nuclear assembly studies. Indeed, the reconstituted structure within the labelled extract can then be isolated and its labelled components analysed easily. Proteins within the extract can be labelled by either of two methods. Eggs prior to use for extract preparation may be injected with a labelled precursor ([^{35}S]methionine, [^{3}H]lysine, or [^{3}H]arginine) (*Protocol 4A*). This method is very efficient, but it is necessary to have a microinjection system. Alternatively, it is possible to prepare a low-speed extract from activated eggs, which is competent for protein synthesis, and to add the labelled precursor to the extract (17) (*Protocol 4B*). It is easier, in this case, to process a greater amount of material, but the specific activity of the extract is usually lower. The rate of protein synthesis is found to be around 30 μg protein/ml extract (17) in the conditions described in *Protocol 2*. Exogenous RNA can be added and new protein synthesized (14).

Protocol 4. Labelling proteins in the egg extract

Equipment and reagents

- Dejellied eggs (*Protocol 1*)
- Labelled precursors: [^{35}S]methionine, [^{3}H]lysine, and [^{3}H]arginine (Amersham; sp. act. 800 Ci/mmol) 10 mCi/ml
- Extraction buffers, tubes, centrifuge rotor (*Protocol 3*)
- Injection setting (A simple system is a Drummond nanoject set on a Brinkman micromanipulator and a stereomicroscope; we use a type M3Z from Leica.)
- 1 × Barth, Ficoll 4%

Protocol 4. *Continued*

A. *Injection of labelled precursors into the eggs*

1. Inject the eggs with a 50 nl solution of labelled precursor (5 μG/μl) in the injection buffer. The injection will cause the activation of the eggs.
2. Transfer the injected eggs to a Petri dish containing Barth–Ficoll mix, and incubate for 2 h at 22 °C.
3. Sort the injected eggs and discard those that have turned white or swollen.
4. Proceed for extract preparation (LSE) as in *Protocol 2*.
5. For HSE continue as in *Protocol 3* or use a small-scale method for a limited number of eggs. In the latter case, centrifuge the low-speed extract for 30 min in an Airfuge (Beckman) at 20 p.s.i. (150 000 *g*).

B. *Direct labelling in the low-speed extract*

1. Prepare the LSE from activated eggs as described in *Protocol 2*.
2. Add the labelled precursors, 1 μl of [^{35}S]methionine 10 μCi to 50 μl of the fresh low-speed extract, and incubate at 22 °C for 2 h.
3. Use the extract directly, taking into account that it has already carried out cell cycles, or freeze as described in *Protocol 2*.
4. For high-speed centrifugation proceed as described in *Protocol 3*.

4. Assembly of nuclear structures

As mentioned earlier, it is possible using *Xenopus* egg extracts and DNA to assemble nuclear structures from the nucleosome (the basic unit of chromatin) into a complete nucleus. The HSE, lacking the membrane precursor, will not reconstitute a nucleus, but the earlier step of chromatin assembly can be followed. The LSE, or the combination of an HSE and the membranous fraction, will support the reconstitution of whole nuclei. The LSE presents the additional property of allowing multiple cell cycles. In the choice of templates, it is important to distinguish purified DNA versus demembranated nuclei. Purified DNA from a variety of sources can be used, e.g. *Xenopus* genomic DNA, bacteriophage lambda DNA, cosmid DNA for a template of long size, various plasmids (derivatives of pUC, pBR etc.), as well as single-stranded DNA (from the M13 virion). The procedure for the preparation of these DNA molecules is described in Maniatis *et al.* (18). Concatemerization of lambda DNA is documented by Newport (19). Demembranated nuclei have been used to follow chromatin remodelling and the dynamics of nuclear organization (20, 21). The most favoured template has been *Xenopus* sperm. However, assays using quiescent somatic cells and tissue culture cells have also been described. Here we will describe, essentially, the process using naked DNA templates.

4.1 Chromatin assembly on naked DNA templates

HSE has been used extensively in these experiments. The extract provides both the histones, assembly factors, and enzymes to mimic the complex assembly process operating *in vivo*. Lacking the membrane precursors, the extract cannot form nuclei and thus cannot replicate double-stranded DNA. The initiation of DNA replication in the *Xenopus* system has been shown to be strictly dependent on the nuclear membrane formation. However, DNA synthesis can be reproduced on a single-stranded DNA template. This property has been very useful in following chromatin assembly coupled to DNA synthesis in the HSE. This process is very rapid and efficient. In contrast, the chromatin assembly on a double-stranded DNA, which is not replicated, is slow. Two chromatin assembly pathways are thus reproduced using one template or the other: the former dependent and the latter independent of DNA synthesis. It has been observed that the kinetics of assembly are very different in the two cases. Furthermore, specific factors may be required for the assembly coupled to DNA synthesis, as demonstrated in a human system.

A convenient model DNA template comes from the phage M13. Double-stranded DNA is obtained from the infected bacteria. Single-stranded DNA is recovered from the phage particules. The circular nature of these DNA templates make them convenient for following the assembly process by analysing the topological changes on DNA (see Chapter 7). For each nucleosome deposited, the constraints arising are relaxed by topoisomerase activity within the extract; hence, the circular DNA after deproteinization displays a number of negative supercoils corresponding to the number of nucleosome formed, which can be analysed by gel electrophoresis to follow the kinetics of assembly. When the reaction is complete it is important to analyse the quality of the reconstituted chromatin. This is achieved using micrococcal nuclease (MNase). MNase cleaves within the linker DNA preferentially, leaving the DNA associated with histone octamer protected. The analysis of the products of digestion reveals whether DNA has been assembled in a regular array of nucleosomes and the nucleosome repeat length. In the HSE, the DNA can be assembled with a regular spacing comparable to that found in the native chromatin from a *Xenopus* embryo. This is a unique feature of the crude system for chromatin reconstitution.

For the assembly reaction, the amount of DNA should be kept low to avoid depleting the endogenous histone pool. We estimate that a standard high-speed extract has a histone concentration around 140 ng/μl, consistent with other estimates found in the literature (3). It should be pointed out that chromatin assembly operates more efficiently on circular DNA compared to linear DNA of small size (22). We describe here the asssembly reaction using a single-stranded DNA template. The assembly reaction is followed by topo-

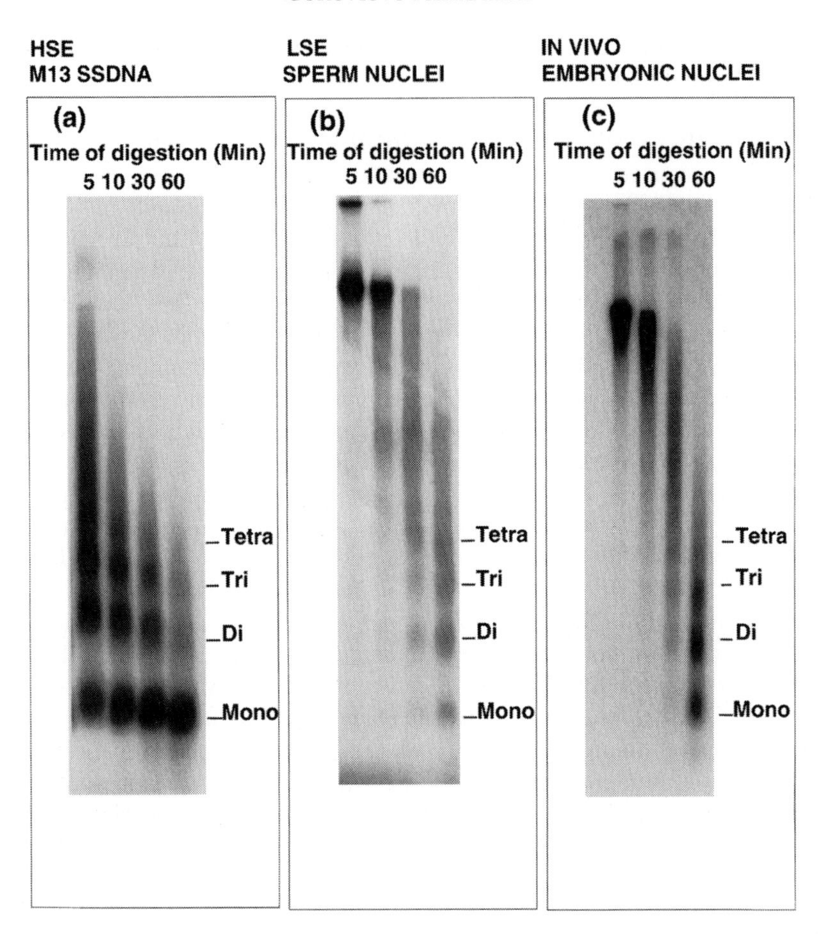

Figure 2. Analysis of chromatin assembly by micrococcal nuclease digestion. The digestion was performed using either a single-stranded DNA template (M13) on which complementary-strand synthesis was performed in an HSE extract in (a) or demembranated sperm nuclei replicated in an LSE extract used fresh in (b). For comparison, digestion was performed on *Xenopus* embryonic nuclei labelled *in vivo* (c). Positions of mono-, di-, tri-, and tetra-nucleosomes are indicated.

logical analysis and MNase digestion (Chapters 7 and 9). Typical results are shown in *Figure 2*.

Protocol 5. The assembly reaction

Reagents

Make all the solutions in extraction buffer to maintain the ionic strength of the reaction.

- HSE
- 100 mM ATP stock. Keep at − 70°C to avoid hydrolysis.
- Single-stranded M13 DNA at 1 mg/ml
- 200 U/μl creatine kinase (Boehringer Mannheim)

- 200 mM $MgCl_2$ stock
- 400 mM creatine phosphate (Boehringer Mannheim)
- [α-^{32}P]dATP (3000 Ci/mmol, 1 mCi/ml; Amersham)

Method

1. Set up a reaction with 25 μl of HSE and 400 ng DNA in a final volume of 30 μl in a 0.5 ml Eppendorf tube. Make it 3 mM ATP, 5 mM $MgCl_2$, 10 U/μl phosphocreatine kinase, 40 mM creatine phosphate, and 1 μl [α-^{32}P]dATP. Complete with extraction buffer to the final volume. Avoid diluting the HSE over 1/3. For a large preparation of chromatin, simply scale up the reaction mix.
2. Incubate the reaction at room temperature (24 °C).
3. To follow the kinetics of assembly take aliquots at the desired time and proceed for supercoiling analysis (Chapters 7 and 9).
4. After 90 min (time usually required for a complete reaction), continue with micrococcal nuclease digestion (Chapter 9).

4.2 Nuclei reconstitution

The use of amphibian extracts to investigate nuclear organization has been pioneered by the work of Lohka and Masui (2) using demembranated sperm nuclei as templates. Newport (19) found that even protein-free bacteriophage DNA can act as a template for nuclear assembly when added to *Xenopus* egg extracts. In the amphibian systems, the components used for nuclear assembly *in vitro* are derived from the excess pool of those made and stored in the egg during oogenesis. Each egg contains enough nuclear component to assemble about 4000 embryonic nuclei. The assembly of nuclear structure occurs spontaneously when DNA is added to a mixture of unassembled nuclear components (19) in a low-speed type of extract. Alternatively, the extract can be fractionated into soluble and particulate fractions by centrifugation as described above. Both fractions are required for nuclear reconstitution. Details on nuclear import and assembly reaction have been presented by Newmeier and Wilson (7). Here we will give a simple standard protocol to follow nuclear formation.

Protocol 6. Nuclear formation and assay by fluorescence microscopy

Equipment and reagents

- Xenopus egg extract: use either LSE (cycling or not) or HSE added with the membrane fraction.
- L-α-Lysophosphatidylcholine (Lysolecithin, Calbiochem)
- Nuclear substrate (lambda DNA stock at 1 mg/ml, demembranated sperm nuclei stock at 50 000 sperm/μl)
- Burker cell-counting plate and 10 × objective

Protocol 6. *Continued*

- ATP-regenerating system: 100 mM ATP stock, 1 M phosphocreatine kinase, 700 U/ml creatine phosphokinase
- Dye buffer: 1 μg/ml Hoechst 33258 (Sigma) or DAPI (Sigma), DHCC (Sigma) dyes, 3.7% formaldehyde, in egg-extract buffer. Store the solution at 4°C; avoid exposure to light. (Optional: add glycerol to 50%. The inclusion of glycerol prevents the slides from drying out, even if they are kept at room temperature overnight.)
- BSA
- Nuclear isolation media (NIM): 0.2 M sucrose, 2.5 mM $MgCl_2$, 10 mM K-Hepes, pH 7.5
- Glycerol mix: 7 ml (NIM + 0.4% BSA) and 3 ml ultra-pure glycerol
- 18 × 18 mm coverslips and microscope slides
- Fluorescence filter set for Hoechst staining
- Phase-contrast microscope
- Nail varnish

A. *Preparation of demembranated sperm nuclei*

All methods involve the use of L-α-lysophosphatidylcholine (2, 13–15, 23, 24).

1. Dissect the testes from 1 male frog (prime a few days before, using 25U PMSG).
2. Trim off the fat, roll on Kleenex paper to remove blood, and rinse in NIM.
3. Mince the testes in 1 ml NIM using a razor blade, then transfer to a 1 ml potter.
4. Perform the subsequent steps at 4°C.
5. Spin for 10 sec in a clinical centrifuge (about 1000 *g*) and collect the sperm in the supernatant.
6. Back-extract the pellet with 300 μl NIM.
7. Spin for 10 min at 1000 *g*.
8. Resuspend the clean top part of the pellet in 500 μl NIM. (The bottom part of the pellet contains somatic cells and erythrocytes.)
9. Pellet for 10 min at 1000 *g* (repeat wash twice).
10. Resuspend in 100 μl NIM and add 300 μl 0.05% lysolecithin. (Dissolve lysolecithin in warm NIM). Leave for 5 min at room temperature, turn the tube upside down every minute, then add 3 volumes of cold NIM + 3% BSA.
11. Pellet for 10 min at 1000 *g*.
12. Wash again with NIM + 3% BSA.
13. Resuspend in 200 μl of glycerol mix.
14. After mixing thoroughly, take an aliquot for counting (a 200-fold dilution is adequate and then use 10 μl).
15. Use the Burker cell-counting plate and 10 × objective and calculate the sperm concentration. The concentration of sperm nuclei/ml = average number of sperm counted per square (n) × 10^4 × dilution factor.

B. *Nuclear assembly assay*

1. Set up the following mix on ice. For a typical nuclear assembly reaction, use 10 μl of LSE or 10 μl of soluble fraction added with 1 μl membrane fraction. Include the ATP regenerating system to give 1 mM ATP, 15 U/μl phosphocreatine kinase, 40 mM creatine phosphate.
2. Add the nuclear substrate, keep the amount between 200 and 2000 nuclear substrates per egg equivalent, assuming that about 0.7 μl corresponds to one egg.
3. Incubate at room temperature for the desired time (30 min to 1 h). If lambda DNA is used with the reconstituted system (membrane fraction and HSE), pre-incubate the first DNA with HSE supplemented with glycogen at 15 mg/ml. Then add the membrane fraction.
4. Withdraw 1 μl of the reaction mix and deposit onto a microscope slide.
5. Add 4 μl of the dye buffer on the top of the drop and cover gently with a 18 × 18 mm coverslip. Hoechst or DAPI will stain DNA and DHCC the membrane components.
6. Examine under a microscope with the appropriate fluorescence filter set using UV fluorescence for Hoechst staining. Use phase-contrast microscopy to visualize membranes.
7. Seal with nail varnish to preserve the slides, and store them at – 20 °C in the dark.

The cytosol and vesicular fractions have been shown to be sufficient to form functional nuclei with the complex chromatin template derived from demembranated *Xenopus* sperm (2). These nuclei form pore-complexes and are competent for transport (7). When naked DNA is used as the template much less is known of the early steps in chromatin assembly and nuclear envelope assembly. When a crude extract is used, long linear DNA substrates acquire nucleosomes and form a highly condensed intermediate (19), and in a multistep manner progress towards the formation of a functional nucleus. Many of these steps are less efficient in a fractionated extract. In the cytosol alone, lambda DNA does not form the condensed intermediate (25); instead, a 'fluffy' non-condensed intermediate is generated. It was recently proven that, in addition to the cytosol and vesicle fraction, glycogen, a component of the gelatinous pellet, participates to promote the formation of functional nuclei (26). This would be required during the formation of a chromatin intermediate.

5. Transcription by RNA polymerase III

There are three distinct polymerases, called I, II, and III, involved in the transcription of the three corresponding classes of eukaryotic genes. All three polymerases were initially found in *Xenopus* oocytes and eggs (4). *In vitro* transcription of the small RNA genes (tRNA and 5S RNA) by RNA polymerase III has been most extensively studied using oocyte extracts or egg extracts (4). These extracts are essentially high-speed extracts as defined previously. Polymerase II-dependent transcription has also been reproduced using a similar type of extracts (27, 28). However, for polymerase II, a most favoured approach has been rather to use the *Xenopus* egg extract to assemble chromatin over a DNA template containing the promoter of choice, going eventually up to the organization of synthetic nuclei (29), and then to perform the transcription assay in a heterologous system. Alternatively, polymerase II transcription can be followed after microinjection into *Xenopus* oocytes (30). Concerning polymerase I-dependent transcription, *Xenopus* egg extracts have not yet been reported as being competent to carry out this transcriptional process *in vitro* .

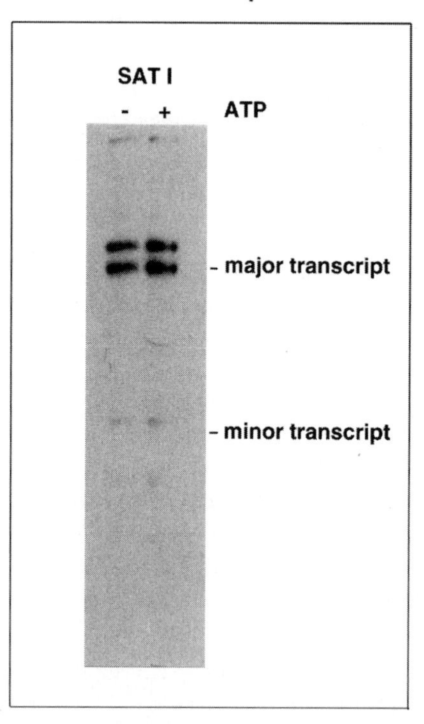

Figure 3. Pol III transcription in the HSE extract. The transcription reaction was as described in *Protocol 7* using as a template the M13 single-stranded form of the derivative containing the Sat I DNA. Comparison is shown for conditions with or without additional ATP. The ^{32}P-labelled transcripts were analysed directly on a sequencing gel. A major transcript is detected in the top part, a minor one in the bottom part. They originate from different start sites.

We will describe here our current method for a transcription assay in the HSE, to analyse polymerase III transcription. As this concerns small RNA genes, it is convenient to follow RNA transcription by a direct labelling method. Genes transcribed by RNA polymerase III belong to two groups. One is represented by genes with regulatory sequences within the transcribed regions, it includes the tRNA genes, satellite I DNA, and 5S RNA genes. The other presents regulatory sequences outside the transcribed region, as the U6 gene encoding (sn) RNA involved in splicing. Regulation of transcription of the 5S RNA gene has been extensively studied using *Xenopus in vitro* systems (4). Oocyte nuclear extracts are sufficient for efficient transcription of the 5S RNA gene (31), whereas the egg extract has to be supplemented with the transcription factor TFIIIA. tRNA or Sat I genes that do not depend on TFIIIA can be transcribed efficiently in both cases. Typical results are shown in *Figure 3*.

Protocol 7. Polymerase III transcription in a high-speed, egg extract

Equipment and reagents

- Equipment and reagents in *Protocol 5*
- DNA template for transcription containing a polymerase III promoter: tRNA, Sat I or 5S RNA gene, in the latter case TFIIIA should be added to the reaction mix[a]. (Nuclei can also be used (see above).)
- Placenta RNAse inhibitor (BRL)
- rNTP (minus C) mix: 20 mM each (Boehringer Mannheim)
- [α-^{32}P]CTP (3000 Ci/mM, 10 mCi/ml; Amersham)
- RNA now (Biogentex)
- Sequencing loading buffer: 80% formamide, 10 mM EDTA, pH 8.0, 1 mg/ml xylene cyanol and Bromophenol blue
- 6% acrylamide, 7 M urea (38:2) in TBE buffer
- 20 × TBE buffer: 216 g Tris-base, 110 g boric acid, 80 ml EDTA 0.5 M, pH 8
- Rotary concentrator (e.g. Speed-Vac)
- TE buffer: 10 mM Tris–HCl, pH 8, 1 mM EDTA
- Photographic film

Method

1. Set up a reaction with 10 μl of HSE and 100 ng DNA template in a final volume of 15 μl in a 0.5 ml Eppendorf tube. Make it 3 mM ATP, 5 mM $MgCl_2$, 10 U/μl phosphocreatine kinase, 40 mM creatine phosphate, 0.5 μl RNase inhibitor, 500 μM rNTP (minus C) mix, and 1 μl [α-^{32}P]CTP. Complete with extraction buffer to the final volume. Avoid diluting the HSE over 1/3.
2. Incubate the reaction for the desired time at 24°C. A 1-h incubation time is sufficient for a good accumulation of transcripts.
3. Stop the reaction by adding 200 μl of RNA now.
4. Homogenize and add 20 μl of chloroform. Mix carefully.
5. Leave on ice for 15 min.
6. Centrifuge at 4°C in an Eppendorf centrifuge for 15 min at 10 000 *g*.

Protocol 7. *Continued*

7. Remove the supernatant.
8. Precipitate the supernatant with an equal volume of isopropanol for 1 h at −20°C.
9. Centrifuge to recover the pellet containing RNA at 10 000*g*, 4°C.
10. Wash once with 500 μl cold 70% (v/v) ethanol.
11. Dry briefly (maximum 3 min) in a rotary concentrator (e.g. Speed-Vac).
12. Resuspend in 50 μl TE buffer and reprecipitate with 200 μl ethanol, for 1 h at −20°C.
13. Centrifuge in an Eppendorf centrifuge for 15 min at 10 000*g*, 4°C.
14. Wash the pellet with 500 μl cold 70% (v/v) ethanol.
15. Dry the pellet (maximum 3 min) in a rotary concentrator.
16. Resuspend in 10 μl sequencing loading buffer.
17. Load 5 μl of the sample on a 6% acrylamide sequencing gel.
18. Run at 30 W until the Bromophenol blue has reached two-thirds of the gel length.
19. Dry the gel, and set up against X-ray film. A 2–3 h exposure is usually sufficient to detect a specific transcription signal. The specific signal is detected according to size migration and by comparison with a mock reaction.

[a] See ref. 32.

6. Replication

Xenopus egg extracts have been used extensively to study DNA replication. High-speed extracts or the soluble fractions are able to reproduce *in vitro* events resembling those occurring at the lagging strand during the elongation process of replication. This has been observed using circular single-stranded molecules as model templates (33). In these extracts, initiation on double-stranded DNA molecules does not occur, as this event is strictly dependent on the presence of membranes to re-form nuclei (13, 15). Low-speed extracts which contain membrane fractions, as well as the mix of membrane fractions together with the soluble fraction can carry out this initiation of replication.

6.1 Complementary-strand synthesis on single-stranded DNA templates

The M13 model system is a useful tool for mimicking events at the lagging strand during replication. This is achieved in the HSE. The single-stranded

DNA isolated from the M13 phage is used as a template for complementary-strand synthesis. To follow DNA synthesis in the reaction one measures the incorporation of [α-^{32}P]dNTP in the newly synthesized DNA.

Protocol 8. Complementary-strand synthesis in a high-speed, egg extract

Equipment and reagents

- Equipment and reagents for *Protocol 5*
- 5% trichloroacetic acid (TCA)
- 5% TCA, 2% sodium pyrophosphate (ppNa). (10 ml per filter)
- Ethanol (technical grade)
- Scintillation counter, vials, and scintillation fluid
- Whatman GF/C filter 2–1 cm. Label each filter with India ink.
- 20 mg/ml proteinase K solution (Sigma)
- Stop-mix solution: 20 mM Tris–HCl, pH 7.5, 1% SDS, 80 mM EDTA
- Tris–glycine buffer: 120 g Tris-576 g glycine in 20 litres H_2O final volume
- 5 × sample buffer: 50% glycerol, 0.5% Bromophenol blue, 0.5% xylene cyanol, 5 mM EDTA
- 1% agarose gel
- Autoradiography equipment and reagents

Method

1. Set up a reaction mix similarly to *Protocol 5*.
2. Incubate the reaction for the desired time (90 min at room temperature (24 °C) is sufficient for a complete reaction).
3. Take two aliquots of 5 μl.
4. Put one aliquot on a GF/C filter for TCA precipitation, allow to adsorb for a few seconds, then drop the filter into a beaker containing 5% TCA, 2% ppNa solution (use 10 ml per filter). After 15 min, change to a fresh 5% TCA solution. Repeat the 5% TCA wash three times.
5. Wash twice with ethanol.
6. Dry the filter.
7. Transfer the filter into a scintillation vials.
8. Add scintillation fluid and count in a scintillation counter.
9. Stop the reaction in the other aliquot by adding an equal volume of the stop-mix solution for gel analysis. Add 2 μl of a 20 mg/ml proteinase K solution and incubate at 37 °C for at least 1 h.
10. Add 1/5 volume of 5 × sample buffer, and run the DNA on a 1% agarose gel in Tris–glycine buffer at 3 V/cm.
11. Dry the gel and visualize DNA by autoradiography.

To determine the incorporation of [α-^{32}P]dATP, the endogenous pool of dATP is determined in the extract. It is usually found to be 50 μM in the high-speed extract (15). Little variation was noticed between different batches of extracts. The efficiency of conversion of the single-stranded template to its double-stranded form is usually close to 100% in the defined working conditions with a good extract.

6.2 Initiation of replication on double-stranded DNA

The initiation of replication within this system is strictly dependent on nuclear formation. As seen previously, low-speed extracts or a reconstituted system

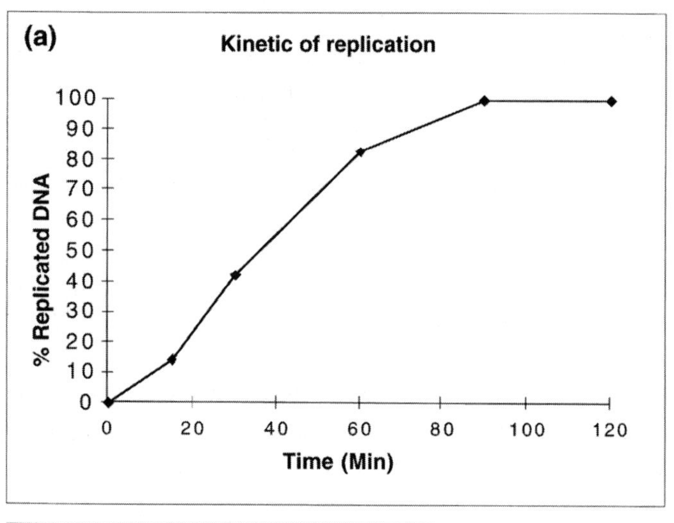

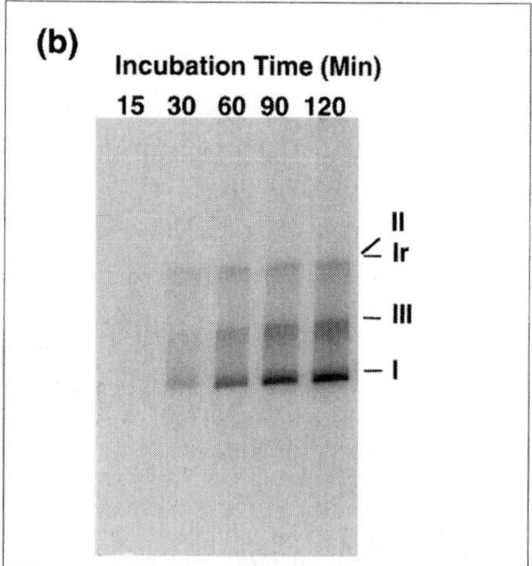

Figure 4. Complementary-strand synthesis in a *Xenopus* HSE. The assay was performed as described in *Protocol 9*. For each time point, as indicated, aliquots were taken and processed for TCA-precipitation or gel analysis. (a) The quantitation is presented as a percentage of input DNA. (b) Shows the analysis of supercoiling on agarose gel. The position of the various forms of replicated DNA generated are indicated: II for nicked double-stranded circular DNA; Ir for relaxed double-stranded circular DNA; III for linear double-stranded DNA; I for supercoiled double-stranded circular DNA.

(HSE supplemented with membrane fraction) can reproduce the nuclear formation and are therefore appropriate for use in studying the initiation of replication. The advantage of the reconstituted extract is that the HSE can easily be immunodepleted of various components to test their individual roles. However, it should be noted that there is a progressive loss of efficiency in the fractionated-reconstituted system which is even more dramatic when the substrate is a naked DNA.

Essentially all the templates used in the nuclear reconstitution assay can be used. The assays for DNA replication are mainly of two types. The first one relies on the incorporation of [α-^{32}P]dNTP into the newly synthesized DNA, which can be followed by gel analysis, or by directly counting the incorporation, as in *Protocol 8*. The other method makes use of fluorescence microscopy to visualize sites of synthesis by the incorporation of BrdUTP or biotinyl-dUTP and revelation with an antiBrdUTP antibody coupled to a fluorescent marker or streptavidin–Texas red (34). Replication domains have been identified by this method in *Xenopus* sperm nuclei replicating in the egg extract (35). Some advantages of the biotin–streptavidin system over the BrdU antigen–antibody are as follows:

(a) The biotin–dUTP incorporation is proportional to DNA synthesis.
(b) There is no requirement for DNA denaturation to facilitate streptavidin binding to biotin as in the antibody–antigen system.
(c) The affinity of the biotin–streptavidin is greater, and there is generally a lower background in these reactions.

Protocol 9. Initiation of replication, assay by incorporation of biotinyl-dUTP and fluorescence microscopy

Equipment and reagents

- Similar to *Protocol 6* with the addition of the following:
- LSE or the reconstituted system (HSE and the membrane fraction)
- Biotinyl-dUTP (Bio-1-dUTP, Enzo diagnostic)
- Streptavidin–Texas red (Amersham)
- polyLysine-coated glass coverslips (12 mm diameter)
- 1 × phosphate-buffered saline (PBS)
- 15 ml Corex tubes with Plexiglass adaptators at the bottom[a]
- 30% glycerol in PBS
- Humidified box
- BSA
- PBS/0.1% Tween
- 50 mM NH_4Cl solution in PBS
- Sodium azide
- Hoechst 33258 diluted in: PBS, 0.05% Tween, 1.5% BSA, 0.2% Na-azide
- RNase A
- Citifluor (Chemical Laboratory, the University of Kent, Canterbury, ref AF1)
- Nail varnish

Method

1. Set up the reaction as in *Protocol 7* and include approximately 10–30 μM biotin-11-dUTP.
2. Incubate at 23°C for the desired time (about 1 h).

Protocol 9. *Continued*

3. Dilute the reaction mixture with 3 vols egg extract buffer, and add the same volume of fixative (4% paraformaldehyde in egg extract buffer) for 10 min.
4. Prepare 15 ml Corex tubes with Plexiglass adaptators at the bottom[a]. Add a glass coverslip to each tube. Fill each tube with 30% glycerol in PBS.
5. Gently layer the fixed samples over the glycerol cushion, and centrifuge at 1500 *g* for 15 min.
6. Retrieve the coverslip from the Corex tube, remove the glycerol buffer by touching the edge of the coverslip onto a paper towel.
7. Place the coverslip in a humidified box.
8. Quench by adding a drop of 50 mM NH_4Cl in PBS on top of the coverslip. Leave for 10 min.
9. Wash with PBS/0.1% Tween. Leave for 1 min.
10. Stain for 30 min with fluorescent streptavidin–Texas red (1/50 dilution) and with the diluted Hoechst 33258 dye to stain DNA. Include RNase A (50 μg/ml) to the stain mixture to eliminate any possible non-specific signal from labelled RNA.
11. Wash three times with PBS, 0.1% Tween. Take 3–4 min for each wash.
12. Mount in Citifluor by putting 5 μl of Citifluor on the microscope slide (not the coverslip). Place the coverslip gently on the top of this drop and let it dry, seal with nail varnish.
13. Examine under conventional or confocal microscope using the appropriate fluorescence mode (*Protocol 6*).

[a] See ref. 36.

Replication sites appear as foci (35). It is possible to follow the distribution of specific proteins involved in the replication process, which involves the use of specific antibodies and subsequent immunodetection on the preparation. PCNA, DNA polymerase alpha, and RPA have been localized using demembranated sperm nuclei as substrate in the reaction (37, 38), as well as specific nucleolar proteins (39). This approach is likely to provide information for the definition of functional domains within a nucleus.

Acknowledgements

I thank Sylvie Milley for her help with the presentation of the manuscript, Michel Bornens and Roy Golsteyn for critical reading.

References

1. Lewin, B. (1994). *Cell*, **79**, 397.
2. Lohka, M. J. and Masui, Y. (1983). *Science*, **220**, 719.
3. Laskey, R. A., Mills, A. D., and Morris, N. R. (1977). *Cell*, **10**, 237.
4. Almouzni, G. and Wolffe, A. P. (1993). *Exp. Cell Res.*, **205**, 1.
5. Leno, G. H. (1991). In Xenopus laevis: *practical uses in cellular and molecular biology* (ed. B. K. Kay and H. B. Peng), Vol. 36, p. 561. Academic Press Inc. Harcourt Brace Jovanovitch Publishers.
6. Wolffe, A. P. and Schild, C. (1991). In Xenopus laevis: *practical uses in cellular and molecular biology* (ed. B. K. Kay and H. B. Peng), Vol. 36, p. 541. Academic Press Inc. Harcourt Brace Jovanovitch Publishers.
7. Newmeier, D. D. and Wilson, K. L. (1991) In Xenopus laevis: *practical uses in cellular and molecular biology* (ed. B. K. Kay and H. B. Peng), Vol. 36, p. 607.
8. Murray, A. (1991) In Xenopus laevis: *practical uses in cellular and molecular biology* (ed. B. K. Kay and H. B. Peng), Vol. 36, p. 581. Academic Press Inc. Harcourt Brace Jovanovitch Publishers.
9. Félix, M-A, Clarke, P. R., Coleman, J.,Verde, F., and Karsenti, E. (1993). In *Cell cycle: a practical approach* (ed. P. Fantes and R. Brooks), p. 253. IRL Press, Oxford.
10. Wu, M. and Gerhart, J. (1991) In Xenopus laevis: *practical uses in cellular and molecular biology* (ed. B. K. Kay and H. B. Peng), Vol. 36, p. 3.
11. Newmeier, D. D., Farshon, D. M., and Reed, J. C. (1994). *Cell*, **79**, 353.
12. Lohka, M. J. and Maller, J. (1985). *J. Cell Biol.*, **101**, 518.
13. Hutchison, C. J., Cox, R., Drepaul, R. S., Gomperts, M., and Ford, C. C. (1987). *EMBO J.*, **6**, 2003.
14. Murray, A. W. and Kirschner, M. W. (1989). *Nature*, **339**, 275.
15. Blow, J. J. and Laskey, R. A. (1986). *Cell*, **47**, 577.
16. Murray, A. W., Solomon, M. J., and Kirschner, M. W. (1989). *Nature*, **339**, 280.
17. Patrick, T. D., Lewer, C, E., and Pain, V. M. (1989). *Development*, **106**, 1.
18. Maniatis, T., Fritsch, E. F., and Sambrook, J. (1982). *Molecular cloning: a laboratory manual.* Cold Spring Harbor Laboratory, Cold Spring Harbor, NY.
19. Newport, J. (1987). *Cell*, **48**, 205.
20. Philpott, A., Leno, G. H., and Laskey, R. A. (1991). *Cell*, **65**, 569.
21. Dasso, M., Dimitrov, S., and Wolffe, A. P. (1994). *Biochemistry*, **91**, 12477.
22. Almouzni, G. and Méchali, M. (1988). *EMBO J.*, **7**, 4355.
23. Gurdon, J. B. (1976). *J. Embryol. Exp. Morphol.*, **36**, 523.
24. Coppock, D. L., Lue, R. A., and Wangh, L. J. (1989). *Dev. Biol.*, **131**, 102.
25. Hirano, T. and Mitchison, T. J. (1991). *J. Cell Biol.*, **115**, 1479.
26. Hartl, P., Olson, E., Dang, T., and Forbes, D. J. (1994). *J. Cell Biol.*, **124**, 235.
27. Tafuri, S. R. and Wolffe, A.P. (1990). *Proc. Natl Acad. Sci. USA*, **87**, 9028.
28. Toyoda, T. and Wolffe, A. P. (1992). *Anal. Biochem.*, **203**, 340.
29. Barton, C. M. and Emerson, B. M. (1994). *Genes Dev.*, **8**, 2453.
30. Almouzni, G. and Wolffe, A. P. (1993). *Genes Dev.*, **7**, 2033.
31. Birkenmeier, E. M., Brown, D. D., and Jordan, E. (1978). *Cell*, **15**, 1077.
32. Almouzni, G., Méchali, M., and Wolffe, A. P. (1990). *EMBO J.*, **9**, 573.
33. Almouzni, G. and Méchali, M. (1988). *EMBO J.*, **7**, 665.
34. Nakamura, H., Morita, T., and Sato, C. (1983). *Exp. Cell Res.*, **165**, 291.

35. Mills, A. D., Blow, J. J., White, J. G., Amos, W. B., Wilcock, D., and Laskey, R. A. (1989). *J. Cell Sci.*, **94**, 471.
36. Evans, L., Mitchison, T. J., and Kirschner, M. W. (1985). *J. Cell Biol.*, **100**, 1185.
37. Hutchison, C. and Kill, I. (1989). *J. Cell Sci.*, **93**, 605.
38. Adachi, Y. and Laemmli, U. K. (1994). *EMBO J.*, **13**, 4153.
39. Bell, P., Dabauvalle, M-C., and Scheer, U. (1992). *J. Cell Biol.*, **118**, 1297.

List of suppliers

5 Prime-3 Prime Inc., 5603 Arapahoe Avenue, Boulder, CO 80303, USA.
Agar Scientific Ltd, 66A Cambridge Road, Stansted, Essex CM24 8DA, UK.
Aldrich Chemical Company, 1001 West Saint Paul Avenue, Milwaukee, WI 53233, USA.
Amersham
Amersham International plc., Lincoln Place, Green End, Aylesbury, Buckinghamshire HP20 2TP, UK.
Amersham Corporation, 2636 South Clearbrook Drive, Arlington Heights, IL 60005, USA.
Amersham, 12, avenue des Tropiques, BP 144, 91144 Les Ulis Cedex, France.
Anderman and Co. Ltd., 145 London Road, Kingston-Upon-Thames, Surrey KT17 7NH, UK.
Applied Biosystems, Kelvin Close, Birchwood Science Park, Warrington, Cheshire WA3 7PB, UK
Bal-Tec AG, FL-9496 Balzers, Lichtenstein.
Baxter Healthcare Corp.
Baxter Healthcare Corp., Salthouse Rd., BrackIIs Industrial Estate, Northampton, NNH OUP, UK.
Baxter Healthcare Corp., Scientific Products, 1430 Waukegan Rd., McGaw Park, IL 60085, USA.
Bayer (Schweiz) AG, Grubenstrasse 6, Postfach, 8045 Zürich, Switzerland.
Beckman Instruments
Beckman Instruments UK Ltd., Progress Road, Sands Industrial Estate, High Wycombe, Buckinghamshire HP12 4JL, UK.
Beckman Instruments Inc., PO Box 3100, 2500 Harbor Boulevard, Fullerton, CA 92634, USA.
Beckman, 52, Chemin des Bourdons, 93220 Gragny,
Beckman, Holderstr. 14, CH-4057 Basel, Switzerland.
Becton Dickinson
Becton Dickinson and Co., Between Towns Road, Cowley, Oxford OX4 3LY, UK.
Becton Dickinson and Co., 2 Bridgewater Lane, Lincoln Park, NJ 07035, USA.

Becton Dickinson, 94–96, rue Victor Hugo, 94200 Ivry, France.
Becton Dickinson AG, Immengasse 7, CH-4056 Basel, Switzerland.

Bio

Bio 101 Inc., c/o Statech Scientific Ltd, 61–63 Dudley Street, Luton, Bedfordshire LU2 0HP, UK.
Bio 101 Inc., PO Box 2284, La Jolla, CA 92038–2284, USA.

Bio-Rad Laboratories

Bio-Rad Laboratories Ltd., Bio-Rad House, Maylands Avenue, Hemel Hempstead HP2 7TD, UK.
Bio-Rad Laboratories, Division Headquarters, 3300 Regatta Boulevard, Richmond, CA 94804, USA.
Bio-Rad, 2200 Wright Avenue, Richmond, CA 94804, USA.
Bio-Rad Laboratories, Life Science Group, 2000 Alfred Nobel Drive, Hercules, CA 94547, USA.

Biogentex Inc., distributed by OZYME.

Boehringer Mannheim

Boehringer Mannheim UK (Diagnostics and Biochemicals) Ltd., Bell Lane, Lewes, East Sussex BN17 1LG, UK.
Boehringer Mannheim Corporation, Biochemical Products, 9115 Hague Road, PO Box 504, Indianopolis, IN 46250–0414, USA.
Boehringer Mannheim Biochemica, GmbH, Sandhofer Str. 116, Postfach 310120, D-6800 Ma 31, Germany.

British Drug Houses (BDH) Ltd., Poole, Dorset, UK.

BRL, Life Technologies (GIBCO-BRL), 8717 Grovemont Circle, PO Box 6009, Gaithers 20884-9980, USA

Calbiochem, 42, rue d'Arthelon, 92190 Meudon, France.

Calbiochem-Novabiochem International, PO Box 12087, La Jolla, CA 92039-2087, USA.

Calbiochem-Novabiochem (UK) Ltd., Boulevard Industrial Park, Padge Road, Beeston, Nottingham NG9 2JR, UK.

Chemical Laboratory, The University of Kent, Canterbury, UK.

Corning Science Products

Corning Science Products, Bibby Sterilin Ltd, Stone Staffordshire, ST15 USA, UK.
Corning Science Products, Enterprise Avenue, Oneonta, NY 13820, USA.

CRBM-CNRS, Elevage de Xénopes, BP 5051, 34033 Montpellier Cedex, France.

Desaga Labortechnik Gmbh, Postfach 101969. D-69009, Heidelberg, Germany.

Difco Laboratories

Difco Laboratories Ltd., PO Box 14B, Central Avenue, West Molesey, Surrey KT8 2SE, UK.
Difco Laboratories, PO Box 331058, Detroit, MI 48232–7058, USA.

Du Pont
Dupont (UK) Ltd., Industrial Products Division, Wedgwood Way, Stevenage, Herts, SG1 4Q, UK.
Du Pont Co. (Biotechnology Systems Division), PO Box 80024, Wilmington, DE 19880–002, USA.

Du Pont/New England Nuclear (NEN) Research Products, 549 Albany Street, Boston, MA 02118, USA.

Dynal Inc.
Dynal Inc., 45 North Station Plaza, Great Neck, NY 1179, USA.
Dynal Inc., N-0212, Oslo, Norway.

Enzo Diagnostics, 60, Executive Blvd, Farming Dale, NY 11735, USA.

European Collection of Animal Cell Culture, Division of Biologics, PHLS Centre for Applied Microbiology and Research, Porton Down, Salisbury, Wilts SP4 0JG, UK.

Falcon (Falcon is a registered trademark of Becton Dickinson and Co.)

Faust Laborbedarf AG, CH-8201 Schaffhausen, Switzerland.

Fisher Scientific Co.
Fisher Scientific UK, Bishop Meadow Road, Loughborough, Leicestershire LE11 0RG, UK.
Fisher Scientific, 711 Forbes Avenue, Pittsburgh, PA 15219–4785, USA.

Flow Laboratories, Woodcock Hill, Harefield Road, Rickmansworth, Hertfordshire WD3 1PQ, UK.

Fluka
Fluka Chemicals Ltd., The Old Brickyard, New Road, Gillingham, Dorset SP8 4JL, UK.
Fluka Corp., 980 South Second Street, Ronkonkoma, NY 11779, USA.
Fluka-Chemie AG, CH-9470, Buchs, Switzerland.

General Electric Medical Systems, PO Box 414, Milwaukee, WI 53201, USA.

Gibco-BRL
Gibco-BRL (*Life Technologies Ltd.*), Trident House, Renfrew Road, Paisley PA3 4EF, UK.
Gibco-BRL (*Life Technologies Inc.*), 3175 Staler Road, Grand Island, NY 14072–0068, USA.
Gibco-BRL (*Life Technologies Inc.*), PO Box 9418, Gaitherburg, MD 20898, USA
Gibco-BRL (*Life Technologies Trading AG*), Uferstrasse 90, Postfach 533, 4019 Basel, Switzerland.

Glen Research Inc., 44901 Falcon Place, Suite 113, Sterling, VA 20166 USA.

Hoeffer Scientific Instruments, 654 Minnesota Street, Box 77387, San Francisco, CA 94107–0387, USA.

Arnold R. Horwell, 73 Maygrove Road, West Hampstead, London NW6 2BP, UK.

Hybaid
Hybaid Ltd, 111–113 Waldegrave Road, Teddington, Middlesex TW11 8LL, UK.

Hybaid, National Labnet Corporation, PO Box 841, Woodbridge, NJ 07095, USA.

HyClone Laboratories, 1725 South HyClone Road, Logan, UT 84321, USA.

ICN Biomedicals Inc., Research Products Division, PO Box 5023, Costa Mesa, CA 92626, USA.

International Biotechnologies Inc., 25 Science Park, New Haven, CT 06535, USA.

Intervet International BV., Boxmeer, Holland.

Invitrogen Corporation

Invitrogen Corporation, 3985 B Sorrenton Valley Building, San Diego, CA 92121, USA.

Invitrogen Corporation c/o British Biotechnology Products Ltd., 4–10 The Quadrant, Barton Lane, Abingdon, OX14 3YS, UK.

Irvine Scientific, 2511 Daimler Street, Santa Ana, CA 92705, USA.

ISCO, **Instrumenten Gesellschaft**, Räffelstr. 32, CH-8045 Zürich, Switzerland.

Kodak: Eastman Fine Chemicals, 343 State Street, Rochester, NY, USA.

Leica, 86, Avenue du 18 Juin 1940, 92563 Rueil Malmaison Cedex, France.

Life Technologies Inc.

Life Technologies Inc., 8451 Helgerman Court, Gaithersburg, MN 20877, USA.

Life Technologies Inc., PO Box 68, Grand Island, NY 14072–0068, USA.

Medicell International Ltd., 239 Liverpool Road, London N1 1LX, UK

Merck

Merck Industries Inc., 5 Skyline Drive, Nawthorne, NY 10532, USA.

Merck, Frankfurter Strasse, 250, Postfach 4119, D-64293, Germany.

Merck (Schweiz) AG, Rüchligstr. 20, CH- 8953 Dietikon, Switzerland.

Micron Separations Inc., 135 Flanders Road, PO Box 1046, Westborough, MA 01581–6046, USA.

Millipore

Millipore (UK) Ltd, The Boulevard, Blackmoor Lane, Watford, Herts WD1 8YW, UK.

Millipore Corp./Biosearch, PO Box 255, 80 Ashby Road, Bedford, MA 01730, USA.

Millipore AG, Chriesbaumstr. 6, CH-8604 Volketswil, Switzerland.

National Scientific Company, 975 Progress Circle, Lawrenceville, GA 30243, USA.

National Biosciences, 3650 Annapolis Lane North, Suite 140, Plymouth, MN 55447, USA.

New England Biolabs (NBL)

New England Biolabs (NBL), 32 Tozer Road, Beverley, MA 01915–5510, USA.

New England Biolabs (NBL), c/o CP Labs Ltd, PO Box 22, Bishops Stortford, Herts CM23 3DH, UK.

Nikon Corporation, Fuji Building, 2–3 Marunouchi 3-chome, Chiyoda-ku, Tokyo, Japan.

Owl Scientific Plastics, PO Box 566, Cambridge, MA 02139, USA.
OZYME, 10, avenue Ampère, 78180 Montigny le Bretonneux, France.
Perkin-Elmer
Perkin-Elmer Ltd., Maxwell Road, Beaconsfield, Bucks HP9 1QA, UK.
Perkin Elmer Ltd., Post Office Lane, Beaconsfield, Bucks HP9 1QA, UK.
Perkin Elmer-Cetus (The Perkin-Elmer Corporation), 761 Main Avenue, Norwalk, CT 06859–0156, USA.
PerSeptive Biosystems UK Ltd., 3 Harforde Court, Foxholes Business Park, John Tate Road, Hertford SG13 7NW, UK
Pharmacia Biosystems
Pharmacia Biosystems Ltd., (Biotechnology Division), Davy Avenue, Knowlhill, Milton Keynes, MK5 8PH, UK.
Pharmacia LKB Biotechnology AB, Björngatan 30, S-75182 Uppsala, Sweden.
Pharmacia Biotech
Pharmacia Biotech, 23 Grosvenor Road, St Albans, Herts AL1 3AW, UK.
Pharmacia Biotech Inc., 800 Centennial Avenue, Piscataway, NJ 08854, USA.
Pharmacia Biotech (*Europe*), Procordia EuroCentre, Rue de la Fuse-e 62, B-1130 Brussels, Belgium.
Pharmacia Biotech (*Schweiz*) AG, Lagerstrasse 14, Postfach, 8600 Dübendorf, Switzerland.
Pierce
Pierce and Warriner (UK) Ltd., 44 Uper Northgate Street, Chester CH1 4EF, UK.
Pierce, 3747 N. Meridian Road, P.O.Box 117, Rockford, IL 61105, USA.
Pierce Europe BV., Socochim SA, Chemin du Trabandan 28, 1006 Lausanne, Switzerland.
Promega
Promega Ltd, Delta House, Enterprise Road, Chilworth Research Centre, Southampton, UK.
Promega Corporation, 2800 Woods Hollow Road, Madison, WI 53711–5399, USA.
Qiagen
Qiagen Inc., c/o Hybaid, 111–113 Waldegrave Road, Teddington, Middlesex, TW11 8LL, UK.
Qiagen Inc., 9259 Eton Avenue, Chatsworth, CA 91311, USA.
Savant Instruments Inc., 221 Park Avenue, Hicksville, NY 11801, USA.
Schleicher and Schuell
Schleicher and Schuell Inc., D-3354 Dassel, Germany.
Schleicher and Schuell Inc., c/o Andermann and Company Ltd.
Schleicher and Schuell Inc., Keene, NH 03431A, USA.
Sefar AG, Moostr. 2, CH-8803 Rüschlikon, Switzerland.
Serva GmbH, Postfach 1505, D-6900 Heidelberg, Germany.

Shandon Scientific Ltd., Chadwick Road, Astmoor, Runcorn, Cheshire WA7 1PR, UK.

Sigma Chemical Company

Sigma Chemical Company (UK), Fancy Road, Poole, Dorset BH17 7NH, UK.

Sigma Chemical Company, 3050 Spruce Street, PO Box 14508, St Louis, MO 63178–9916, USA.

Sorenson BioScience Inc., 6507 South 400 West Street, Salt Lake City, Utah 84107, USA.

Sorvall DuPont Company

Sorvall DuPont Company, Biotechnology Division, PO Box 80022, Wilmington, DE 19880–0022, USA.

Sorvall (DuPont), Digitana AG, Burghaldenstr. 11, CH-8810 Horgen, Switzerland.

Spectra/Por, Socochim SA, Chemin du Trabandan 28, CH-1006 Lausanne, Switzerland.

Stratagene

Stratagene Ltd., Unit 140, Cambridge Innovation Centre, Milton Road, Cambridge CB4 4FG, UK.

Stratagene Inc., 11011 North Torrey Pines Road, La Jolla, CA 92037, USA.

Tetko Inc., 333 S. Highland Ave., Briarcliffe Manor, NJ 10510, USA.

United States Biochemical Corporation (USB), PO Box 22400 Cleveland, OH 44122, USA.

University Marine Biology Station, Millport, Isle of Cumbrae, Scotland.

USA Scientific Plastics, PO Box 3565, Ocala, FL 34478, USA.

Ultra Violet Products Ltd., The Science Park, Milton Road, Cambridge CB4 4FH, UK.

VG BioTech Ltd., Tudor Road, Altrincham, WA14 5RZ, UK.

Wellcome Reagents, Langley Court, Beckenham, Kent BR3 3BS, UK.

Whatman

Whatman International Ltd., Whatman House, St. Leonards Road, 20/20 Maidstone, Kent ME16 0LS, UK

Whatman LabSales Inc., 9 Bridewell Place, Clifton, NJ 07014, USA.

Whatman, c/o Fisher ScientificWorthington Biochemical Corporation, Halls Mill Road Freehold, NJ 07728, USA.

Whatman, E. Merck (Suisse) SA, 19 rue des Epinettes, 1227 Les Acacias-Genève, Switzerland.

Wheaton Science Products

Wheaton Science Products, Zinsser Analytic UK Ltd, Howarth, Stafferton Way, Maidenhead, Berks SL6 1AP, UK.

Wheaton Science Products, 1000 N. 10th Street, Millville, N. 08332–2092, USA.

Worthington Biochemical Corporation, Lorne Laboratories Ltd., PO Box 6, Twyford, Reading RG10 9NL, UK.

Index

Index

Index